School of Oriental and African Studies

INDUSTRIALIZATION AND AGRICULTURAL SURPLUS

Industrialization and Agricultural Surplus

A Comparative Study of Economic Development in Asia

MASSOUD KARSHENAS

OXFORD UNIVERSITY PRESS
1995

Oxford University Press, Walton Street, Oxford OX2 6DP
Oxford New York
Athens Auckland Bangkok Bombay
Calcutta Cape Town Dar es Salaam Delhi
Florence Hong Kong Istanbul Karachi
Kuala Lumpur Madras Madrid Melbourne
Mexico City Nairobi Paris Singapore
Taipei Tokyo Toronto
and associated companies in
Berlin Ibadan

Oxford is a trade mark of Oxford University Press

Published in the United States
by Oxford University Press Inc., New York

British Library Cataloguing in Publication Data
Data available

Library of Congress Cataloging in Publication Data
Karshenas, Massoud.
Industrialization and agricultural surplus : a comparative study of economic development in Asia / Massoud Karshenas.
Includes bibliographical references.
1. Industrialization—Asia—Case studies. 2. Industries—Asia—Finance—Case studies. 3. Surplus agricultural commodities—Asia—Case studies. 4. Input-output analysis—Asia. I. Title.
HC412.K324 1994 338.95—dc20 94-29842
ISBN 0-19-713610-9

1 3 5 7 9 10 8 6 4 2

Typeset by Create Publishing Services Ltd, Bath, Avon
Printed in Great Britain
on acid-free paper by
Biddles Ltd, Guildford and King's Lynn

Preface

Traditionally, the financing role of agriculture in the early stages of industrialization has been the main focus of research on intersectoral resource flows in the economics literature. In the present study, in addition to addressing the old debate on the role of agricultural surplus in financing industrialization, I have tried to elaborate some of the broader implications of the empirical findings which may be of interest to the more contemporary debates on trade and industrial policy.

The observed patterns of intersectoral resource flows are of significant value in assessing the different theoretical standpoints on the long-term processes of growth and structural change, as different theories invariably involve particular assumptions regarding the patterns and mechanisms of intersectoral resource flows. My own interest in the study of intersectoral resource flows originated in an earlier attempt at such an empirical assessment—related to the 'Dutch Disease' theory and its predictions regarding structural change in oil-exporting economies following an oil price boom. Given the complex processes involved in structural change, the attempts in the literature at the verification of the predictions of theory in the light of the empirical evidence on sectoral growth performances had not been very informative. In my study of the impact of the oil price boom on structural change in Iran, I found it more fruitful to examine the intermediate mechanisms of intersectoral resource transfer which, according to theory, are expected to give rise to the predicted patterns of structural change. The chapter on Iran in the present book is in fact based on the findings of this earlier research. A test of such mediating mechanisms posited by different theories is clearly more illuminating than a casual empiricist approach which compares final outcomes of complex economic processes, for example in the form of sectoral growth performances, with predictions of different theories. The former approach allows an analysis of the underlying mechanisms of structural change postulated by theory, and it can be helpful in illuminating the points of weakness and strength of different theories, thus pointing the way towards their modification and extension in the light of the empirical evidence. I hope that in this respect the comparative study of intersectoral resource flows in the present book will be of value to other researchers.

This book is also based on some earlier research papers published by the International Labour Office (ILO) (Karshenas, 1989, 1993). I am grateful to Samir Radwan and Ajit Ghose of the Rural Employment Branch of the ILO for their encouragement, and gratefully acknowledge the ILO's financial support for that research project. For the purpose of the present book, various ideas expressed in the earlier research paper have been extended and elaborated, and the estimates of intersectoral resource flows for all the countries (with the exception of Taiwan) have been updated in the light of new empirical evidence and the comments I have received since. The new estimates, however, provide further support to the conclusions arrived at in the earlier papers.

For this book I am most of all indebted to my colleagues in the Economics Department at SOAS, for their constant support and encouragement throughout, and particularly to my head of department Terry Byres who first suggested the idea of the book and graciously provided me with the time to complete the book during the Spring Term of 1992. I wish particularly to express my deepest gratitude to Laurence Harris and Michael Hodd for carrying most of the burden of my teaching and supervision load during that term.

I am grateful to the organizers of, and participants in, various conferences and seminars, who have contributed to making this a better book by their constructive criticisms and comments, and who are too numerous to be named individually. Different chapters of the book have been read and commented upon by various people whose help is greatly appreciated. Francesca Bettio, Amit Bhaduri, Ajit Bhalla, Krishna Bharadwaj, Terry Byres, C. P. Chandrasekhar, Naomi Collett, Nigel Crook, Graham Dyer, Michael Ellman, Ben Fine, Ajit Ghose, Hassan Hakimian, Laurance Harris, Christopher Howe, P. K. J. Kikuchi, Andy McKay, Val Moghadam, Utsa Patnaik, Samir Radwan, Nazy Sedaghat, Manik Sen, John Sender, Hans Singer, Ajit Singh, Ravi Srivastava, Hamid Tabatabai, Oktar Turel, Nalini Vittal, and John Weeks read various chapters or all of the book and provided valuable comments towards refining and shaping the ideas at various stages of the research. I am grateful to them all. I am also greatly indebted to Nalini Vittal for her excellent research assistance, and to Nicola Pike for her meticulous copy editing of the final draft. No organization or person other than the author is responsible for the opinions expressed in the book.

Contents

List of Figures

List of Tables

Introduction

This book investigates the direction and pattern of intersectoral resource flows in the process of development on the basis of empirical evidence from a number of country case studies. In the economic development literature it is often assumed that growth of agrarian economies, at least in the early stages, requires a net 'surplus' transfer from the agricultural sector in order to maintain high enough rates of industrial investment. This proposition may seem like a truism, considering that the growth process is necessarily accompanied by a differentiation in demand and increased division of labour in the economy. Further reflection, however, reveals that there is a variety of definitions of the notion of 'agricultural surplus' in the literature, and that there exists a divergence of opinion amongst different authors regarding the interpretation of the optimum magnitude and direction of resource flows and the appropriate transfer mechanisms. This variety of interpretations arises from the differences in the theoretical perspectives as well as the specific empirical conditions which furnish the underlying assumptions of different theories. Under these conditions, what one finds surprising is the relatively meagre attention that this issue has received in empirical research.

Of course, the past few decades have witnessed the growth of a voluminous empirical literature on different aspects of structural change in the development process. This literature, however, has been mainly concerned with the identification of regularities in structural change in the process of development through cross-section and time series empirical analysis.[1] Though the literature has provided valuable insights into the patterns of structural change which accompany economic development in the long run, empirical work on the mechanisms and processes of intersectoral resource transfers which give rise to such structural transformations has remained scant.[2] This state of affairs is very well portrayed in the opening remarks of a recent literature survey on the patterns of structural change:

> Analysis of structure appears in two variants. The first, and more recent, is concerned with the functioning of economies, their markets, institutions, mechanisms for allocating resources, income generation and its distribution, etc. This is primarily a micro approach, solidly anchored in economic theory with little emphasis on economic history or long-run processes of structural change. In the second variant, economic development is seen as an interrelated set of long-run processes of structural transformation that accompany growth ... This is essentially a comparative approach deriving its information from the historical

[1] See e.g. Kuznets (1966, 1971, 1979), Chenery (1960), Chenery and Syrquin (1975), and Syrquin (1988).

[2] Compared to the voluminous empirical literature on the patterns of structural change in the growth process, comparative historical analysis of the intersectoral resource flows is confined to a few studies: namely, Ishikawa (1967*a*, 1988), Mody *et al.* (1985), and Karshenas (1989).

evolution of the advanced economies and from inter-country associations of structural change and growth. (Syrquin, 1988, p. 205)

Neglect of the causal mechanisms and resource flow processes underlying long-term structural change in the empirical literature has left a considerable gap in our understanding of the development process, which economic theory on its own is not equipped to fill. It has also contributed to the ongoing controversy on development policy, particularly industrialization policy, which essentially is not resolvable at the theoretical level.

Consider, for example, the neoclassical parable of an economy in full employment competitive equilibrium. Industrial policies which seek to accelerate the growth of a particular sector of the economy beyond the market-determined rate can be shown to be inefficient and, in the long run, counter-productive. In this way the industrialization problems of the developing countries are often attributed to the 'misallocation' of resources between different sectors or activities as a result of 'distortionary' policy interventions by the government. However, as soon as the assumptions of full employment, perfectly competitive and complete markets, and fixed technology are reconsidered, as seems fit in the context of a developing economy, the theory loses its predictive and prescriptive power.

With alternative assumptions regarding market structure, economies of scale and learning, externalities, and information theory considerations, recent developments from within mainstream neoclassical economics have reassessed the role of the market in efficient resource allocation and reasserted the need for industrial policy.[3] Similarly, the 'structuralist' school, though beginning from a different methodological starting-point, has also advocated active industrial policies of various types by essentially questioning the realism of the assumptions of the neoclassical competitive equilibrium paradigm in relation to the functioning and structure of the underdeveloped economies.[4] Given the variety of theoretical models which can be constructed with different underlying assumptions as to the structure and functioning of the resource allocation and utilization mechanisms in a developing economy, economic theory alone is clearly inadequate in appraising the conflicting policy prescriptions which may be derived on the basis of different theories. Policy prescriptions must, in the end, rely on empirical appraisals.

Empirical verification of different theories may be conducted at a variety of levels. At one level, one may appraise the realism of the assumptions of different theories. However, given that such assumptions do not normally take the form of testable propositions, empirical verification at this level has not contributed much to the formation of a consensus. A second level of empirical verification, which, with the growing availability of data, has become increasingly popular in recent years, is to assess the adequacy of different theories in relation to the accuracy of their predictions. This normally takes the form of appraising the success or failure

[3] See e.g. Stiglitz (1989) and Helleiner (ed.) (1992).

[4] See e.g. Diamond (1978) and Taylor (1983).

of the alternative policies prescribed by different theories, by comparing policy-on/policy-off situations either in one country or in a number of countries.[5] But, even where these comparisons yield apparently unequivocal results, the relevant causal mechanisms may not have been identified because of the multiplicity of factors involved. Not surprisingly, empirical tests of this nature are seldom conclusive.

A third option is to focus directly on the causal relationships involved in the process of structural change as postulated by different theories. An example may help to clarify what is involved. Take the policy of import substitution. From a traditional neoclassical perspective, it may be argued that import substitution hampers growth by distorting terms of trade, specifically turning them against agriculture. Terms of trade effects are thus the key mechanisms governing structural change, to be focused on in the evaluation of import substitution policies. Before passing judgement, however, a number of related questions have to be addressed. Were there significant terms of trade effects? If so, in what direction and with what consequences? Was the agricultural sector adversely affected by the terms of trade effect? Were there other mechanisms of resource flow at work to produce this outcome? And so on. Underlying these questions is the fact that, in an economy with under-utilized resources and with technological progress, even the predicted relative price effect of import substitution policy need not necessarily hold in the long run.

The empirical investigations in this book are addressed to questions of this nature, which in general relate to resource flow mechanisms underlying long-term structural change. These mechanisms provide a bridge between static micro theory and long-run structural analysis. An underlying question which will be addressed is the extent to which the problems of economic development and structural change in the countries under study are associated with misallocation of resources between sectors or activities, as emphasized by conventional economic theory. It is found that, in a long-term perspective, the policy emphasis may have to be put elsewhere. Production efficiency and productivity gains through fuller and better utilization of resources *within* the sectors seem to be of more crucial significance than the allocation of a given amount of resources *between* the sectors, which has been the main preoccupation of economic theory.[6]

[5] This type of exercise is particularly common in the recent assessments of the structural adjustment policies followed by different developing countries (see e.g. Balassa, 1988; World Bank, 1988, 1990). Other prominent examples related to trade and industrialization policies are Agarwala (1983), World Bank (1983), and Aghazadeh and Evans (1985).

[6] The efficiency of resource use within broad sectors such as agriculture and industry could itself be decomposed into allocative inefficiency between sub-branches and various types of production inefficiency, e.g. X-inefficiency *à la* Leibenstein (1966, 1978), and different kinds of technical inefficiency. The separation of these different sources of inefficiency requires a detailed micro-study which falls beyond the confines of the present book. Since we are here mainly concerned with a broad sectoral level of analysis, particularly of the interaction between industry and agriculture, we do not in general distinguish between these different sources of intrasectoral inefficiency.

The book is organized in two parts. In Part I (Chapters 1, 2, and 3) we discuss the theoretical and methodological issues related to intersectoral resource flows. Chapter 1 discusses the different theoretical perspectives on the role of intersectoral resource flows in economic development, particularly the financing role of agriculture in industrial accumulation. An attempt is made to reconcile the conflicting views of different theories by classifying them according to what each theory assumes about the structure of the underdeveloped economy and the binding constraints on industrial accumulation.

In Chapter 2 we introduce a new accounting framework for intersectoral resource flows which acts as an aid both to measurement and to a more precise definition of the different notions of agricultural surplus discussed in the literature. Such a framework is also necessary for the comparison of the country case studies in Part II, where we have to reconcile resource flow measurements for different countries which are often made with different methodologies. The decomposition of intersectoral resource flows derived from the accounting framework also helps in the analysis of the determinants of resource flows, as discussed in Chapter 3.

In Part II of the book we examine the pattern and processes of intersectoral resource flows for five countries during crucial stages of their development experience. The countries are China, India, Iran, Japan, and Taiwan. Although the number of case studies has had to be limited for the sake of manageability, the sample of economies considered is still wide-ranging enough in terms of initial conditions, resource availabilities, and development strategies to allow useful lessons to be learnt in a comparative historical study of this nature. In effect, each country in this sample can be taken as representative of a distinct class of economies with similar characteristics. Taiwan belongs to the group of newly industrializing economies in which, starting from a resource-poor agrarian economy, relative success in the modernization of agriculture and diversification of the industrial base has been achieved. Iran is an example of a resource-rich oil-exporting economy which, during the 1960s and the 1970s, benefited from rapid growth of foreign exchange revenues from the oil sector. China and India have various similarities in terms of initial conditions and economic structure, but the differences in their development strategies allow instructive comparisons with regard to the role of policy in the resource transfer processes. Pre-war Japan is often quoted as a classic case of agriculture-financed industrialization in the development literature, with possible lessons for present-day developing economies.

However, to be able to draw useful conclusions from the comparison of different country experiences, it is important to pay attention to the similarities and differences in the initial conditions and development strategies which shaped the experience of resource transfer in each. In Chapter 4 we provide an overview of the initial conditions in terms of natural resource endowments, the productivity of resource use as determined by technological and institutional conditions, and the initial capital stock, in the five countries in our sample. This provides an overall comparative background for the study of the experience of intersectoral resource flows in each of the countries in the following five chapters.

Chapters 5 to 9 examine the patterns of intersectoral resource flows in each country, analysing the main factors which influenced the observed patterns of resource flows in the long run. Chapter 10 investigates the implications of different patterns of intersectoral resource flow for economic growth, income distribution, and poverty in a comparative study of the five countries in our sample. An overview of the conclusions of the country study chapters and some of the broader theoretical and policy implications of the results are discussed in Chapter 11.

PART ONE

Theoretical and Methodological Issues

1

Development Theory and Intersectoral Resource Flows

1.1 Introduction

In the development literature the discussion of the role of intersectoral resource flows in economic development has been closely tied to the issue of appropriate development strategies in agrarian economies at the early stages of industrialization. A central question in different development strategies has been the identification of the leading sectors in economic development and the formulation of broad policy frameworks to bolster the growth of the leading sector in an attempt to accelerate the transition process. At this level of analysis, the issue of intersectoral resource flows has often revolved around the role of agriculture in generating or releasing resources for rapid industrial accumulation. The present chapter is mainly concerned with this general theoretical discussion of the role of 'agricultural surplus' in economic development. The more specific questions related to the various social forms of agricultural surplus and the different mechanisms and determinants of surplus flow will be discussed in the following two chapters. The issue of intersectoral resource flows between agriculture and other sectors in the development literature has normally been addressed in terms of the 'contribution' of agriculture to economic development at the early stages of industrialization. This contribution can take different forms, which, following Kuznets (1964), can be enumerated as follows. Agricultural output constitutes an important item of wage goods in the early stages of development, and the growth of agricultural marketed surplus is normally an important precondition for non-inflationary growth of employment in other sectors of the economy. In the early stages of industrialization, when comparative advantages in international trade are inevitably determined by natural resource endowments, agricultural marketed surplus may also make an important foreign exchange contribution to the growth process. The degree of market participation of the agricultural sector is also important on the demand side, as, at least in the early stages of development, it dictates the extent and the structure of the home market. The agricultural sector may also make a factor contribution to the growth process. This could take the form of a net transfer of capital funds for investment in other sectors of the economy. It could also take place through a transfer of labour from the agricultural sector to the other sectors of the economy. One measure of this latter contribution may be the resources embodied in the adult migrant labour force, provided by the agricultural sector during the village life of the labourer.

Of these different functions, the factor contribution of agriculture in the form of capital funds has assumed a central place in the policy debate in development economics. A more precise definition of this contribution and its different constituent elements will be given in the next chapter. Here we may refer to it as the net savings or surplus contribution of agriculture to capital formation in the rest of the economy. One of the controversial aspects of the early debate in the development literature was the extent to which agriculture historically has provided capital funds for industrial accumulation in the developed countries at the early stages of their industrialization, and the desirability and possibility of policy intervention to accelerate the rate of this accumulation by manipulating the direction and magnitude of the agricultural surplus flow in underdeveloped countries. Given that capital formation was perceived to be central to the growth of different sectors, even within theoretical standpoints where other functions of agriculture were given prominence – e.g. its marketed surplus contribution, as in the 'wage fund' theories (see the next section) – the policy debate always seemed to gravitate around the issue of agricultural surplus transfer.

Such a focus on intersectoral resource flows, particularly on the role of agricultural surplus, however, should not make this book of interest solely to the historian of development thought. Intersectoral resource flows also have direct relevance to the recent policy debate on industrialization and structural adjustment in the present-day developing countries. For example, one of the main criticisms levelled in the recent literature against the import substitution policies adopted in developing countries in the post-war period has been the adverse effects of such policies on the agricultural sector, allegedly by diverting investment funds from the sector. This in turn, it is argued, has stifled industrial growth and brought expansion to a halt in these economies. An important aspect of the structural adjustment policies advocated for the developing countries since the 1980s is in effect an attempt to correct this 'misallocation' of resources. Clearly, the issue of intersectoral resource flows, and particularly the agricultural surplus flow, is of direct relevance to this type of argument.

The different contributions of the agricultural sector to the growth process are obviously interdependent and cannot be analysed in isolation. In particular, the savings contribution of agriculture, as we shall see in the next chapter, is the net outcome of various real and financial flows, and hence is affected by all other aspects of agriculture/non-agriculture interdependences. Nevertheless, it is helpful for theoretical clarity to make a conceptual distinction between these different contributions of agriculture. We argue here that what is perceived to be the contribution of agriculture to economic growth within different theoretical perspectives depends on what is assumed to be the main constraint to industrial accumulation and growth. We shall therefore use this as an organizing principle in discussing the different theoretical perspectives on the role of agricultural surplus in the industrialization process. The main tasks of this chapter will be to discuss the different theoretical perspectives on the role of agricultural surplus in economic development and to identify the explicit or implicit assumptions that each theory makes about the structure of the underdeveloped economy which it presupposes. This latter task of identification of the assumptions of the theories is particularly important, as it provides the key to understanding the diverse and often opposing views expressed in the literature on the role of agricultural surplus. It also helps to clarify the differences between the views expressed on intersectoral resource flows in conventional development theories as compared to the neoclassical orthodoxy which has increased its influence in development economics in the past two decades.

The chapter is organized in the following way. In Section 1.2 we discuss the different theoretical perspectives on the role of agricultural surplus in the development process. A distinction is made between the views expressed in conventional development economics, where resource flows are discussed in the context of the transition from a predominantly agrarian, surplus labour economy to an industrialized economy, and those of neoclassical economics, where competitive general equilibrium models normally furnish the context. In the development literature we further distinguish between savings-constrained, food-constrained, and demand-constrained models of industrial accumulation. In Section 1.3 we examine the relation between technological progress in different sectors and intersectoral resource flows. A brief summary and concluding remarks are provided in Section 1.4.

1.2 A Classification of Theoretical Approaches

Two basic assumptions underpin the views of conventional development economics with regard to the role of intersectoral resource flows in economic development.[1] The first relates to the special position of the industrial sector, particularly modern manufacturing, in the development process. This special position derives from the existence of dynamic and static economies of scale and the special role of manufacturing in inducing technological progress in other sectors. We shall return to these issues in more detail at the end of this chapter, but for brevity we shall refer to them here as the dynamic economies or externalities associated with industrialization. The existence of these dynamic economies, believed to give a special position to the manufacturing sector as the 'engine of growth' of the economy, has played a key role in the discussion of intersectoral resource flows in the development literature. The second assumption, rather than

[1] By conventional development economics, we refer here to the body of literature which developed in the post-war period, initiated and highly influenced by the early ideas of economists such as Rosenstein-Rodan, Nurkse, Lewis, Hirschman, Prebisch, and Singer, amongst others. Other influential contributors to arrive at a similar perspective following in the tradition of the classical political economists were Joan Robinson, Kaldor, and Kalecki.

referring to some intrinsic characteristic of certain economic activities, is more related to the observed characteristics of underdeveloped economies. These economies often exhibit persistent disequilibria, particularly visible in rural labour markets with the existence of substantial surplus labour. Though the surplus labour assumption is often made explicit and figures more prominently in the dual economy models which have been used to discuss the intersectoral resource flows, the dynamic economies assumption is of no lesser importance.[2] In fact, it is mainly because of this latter assumption that a sectoral distinction between agriculture and industry is useful. And in the literature, where industrialization is often taken as synonymous with economic development, dynamic economies associated with manufacturing are implicitly assumed. The dynamic interactions between industry and agriculture in the process of development, most notably visible in technological change in agriculture, also signify the fact that it would be a mistake to regard the financial surplus or food surplus provided by the agricultural sector as the contributions of that sector *per se*. Throughout this book we shall refer to these contributions of the agricultural sector, but bearing in mind that they are as much influenced by the interdependent relations between agriculture and other sectors as the conditions of production within agriculture.[3]

To highlight the significance of these assumptions, it would be useful to contrast them with those of the neoclassical orthodoxy. In contrast to the assumption of persistent disequilibria that is found in conventional development economics, the neoclassical theories, with their common substitution and continuity assumptions, rely on the price mechanism to bring about the necessary supply adjustments in the growth process. In a neoclassical model all resources simultaneously and continuously act as a bottleneck to growth along the equilibrium growth path. Furthermore, dynamic effects such as economies of scale in the non-agricultural sector are normally treated as exceptions rather than the rule in such a framework, and they are subject to selective policy intervention rather than general policies at a sectoral level.[4] Within this framework, therefore, sectoral distinctions between agriculture and non-agriculture lose their significance, and the analysis, particularly of policy, is shifted to the project or micro level. Given the assumed perfect mobility of labour and capital, any shortages or slacks which result in deviations from the equilibrium growth path are believed to be removed by the operation of the market

[2] The absorption of agricultural surplus labour in the industrial sector itself can be taken as another example of dynamic economies associated with industrial growth, as it automatically increases the productivity of labour in agriculture. It is in fact this productivity increase which allows dual economy models such as that provided by Lewis (1954) to dispense with the need to assume increasing returns in manufacturing. In the absence of such productivity increases, the industrialization process would come to a halt as a result of the internal terms of trade moving against industry.

[3] This point is recognized by most of the economists who refer to the 'contributions' of agriculture to economic growth, and certainly by Kuznets (1964), who notes the different contributions of agriculture, but it is discussed at length in Millar (1970).

[4] The main problem, of course, is that the assumptions of perfect competition and increasing returns would be incompatible. It is only in the most recent literature that increasing returns have been

mechanism. For example, any shortages of food supplies to the non-agricultural sector resulting from the marketed surplus lagging behind the demand for food would push up agricultural prices and profitability of investment in the agricultural sector, leading to a shift of resources to that sector and the restoration of equilibrium in the food market. In an open economy, there may be an increase in food imports and greater concentration of resources in export-oriented industries. Thus the main thrust of policy recommendation at a macro level is the removal of barriers to the mobility of factors of production and correction of instances of market imperfection. Within such a framework, sectoral outputs become substitutes in the sense that any attempt to accelerate the rate of growth of output of one sector has to be at the expense of other sectors. For example, attempts to accelerate the rate of growth of the non-agricultural sector through protective industrial policies would of necessity lead to a contraction of the agricultural output, as, under the conditions of full employment and non-increasing returns, such policies would necessarily involve a shift of resources from the unprotected to the protected sector.

The assumptions of surplus labour in agriculture and dynamic economies in the non-agricultural sector, on the other hand, give rise to a dynamic complementarity in the discussion of intersectoral relations in conventional development economics which is absent from the competitive equilibrium models of neoclassical economics. We shall discuss the significance of these dynamic complementarities at greater length in Section 1.3. The existence of persistent disequilibria as assumed by conventional development theory implies that industrial accumulation at any particular time could be constrained by specific resources which, given the institutional and technological rigidities in the less developed economies, cannot be resolved through the usual neoclassical substitution mechanisms. These rigidities are sometimes referred to as 'structural' in the sense that they set limits to, and condition, both the field of effective economic intervention by the government and the operation of the market mechanism. Policy recommendations within this framework are made with explicit reference to such constraints. In this literature the major scarcities which confront industrial accumulation in the developing countries have been categorized into three major groups: (*a*) the supply of human resources; (*b*) the supply of domestic savings; and (*c*) the supply of key goods and services. In principle, bottlenecks in the supply of key goods and services – such as capital goods, food, raw materials, and intermediate products – could be overcome through imports. This is not feasible, however, if the rate of growth of foreign exchange earnings is also constrained by obstacles to the diversification and expansion of exports. This also suggests a further possible constraint to industrial accumulation, i.e. the limitation of the domestic market and lack of effective demand. In this literature the timing and direction of agriculture–industry surplus

introduced in models with imperfect competition, but such models have had hardly any impact on mainstream neoclassical thinking in relation to the developing countries (see e.g. Krugman, 1990; Murphy *et al.*, 1989).

transfers, as well as the definition of the relevant concept of agricultural surplus, depend on which of the above constraints is assumed to be binding.

The Savings Constraint

In that part of the literature where the savings constraint is assumed to be binding, accelerated industrial accumulation calls for a net transfer of resources from agriculture to other sectors of the economy. Here, agriculture makes a factor contribution in the form of a net financial surplus for investment in the non-agricultural sectors (see e.g. Johnston and Mellor, 1961; Fei and Ranis, 1964, 1966; Owen, 1966). For example, according to Fei and Ranis (1964, p. 31): 'in a dualistic type of underdeveloped economy with a large subsistence agricultural sector, this sector must serve as a primary basis for the expansion of the economy. For this reason, when the economy gathers momentum it is likely that ... [agricultural savings] will constitute a major source of the economy's investment fund, dwarfing the savings of the industrial sector.' For this argument to form a basis for policy intervention, further assumptions are necessary – in addition to the savings constraint assumption. First is the existence of an unrealized potential for increasing savings in the agricultural sector. Such a possibility can exist on a major scale in backward agrarian economies with a low level of financial development in the countryside and with a large share of agricultural incomes being consumed by a parasitic landlord class or rich peasant proprietors. Further assumptions are higher returns on investment in other sectors, and/or the existence of economies of scale in the non-agricultural sectors. It is important to note that, under the conditions of increasing returns in the non-agricultural sector, the prevailing rates of return on investment in non-agriculture need not be higher than in agriculture to warrant a transfer of surplus from the latter to the former. With the existence of economies of scale, the rate of return on investment in industry can increase substantially if a concerted investment effort in industry is made. This was the basis of the idea of the 'big push' put forward by Rosenstein-Rodan (1943). If such a 'big push' happens to be constrained initially by the availability of investment funds, then, in a dual economy context, it may form the basis for the argument of a savings transfer from agriculture as in the sources quoted above.

The savings constraint approach may be traced back to the work of the nineteenth-century classical political economists, and was also influential in the Soviet industrialization debate in the 1920s. For example, Ricardo's arguments in favour of the repeal of the Corn Laws was partly based on the assumption of unproductive utilization of rents in the agricultural sector (unrealized potential for increasing agricultural savings) and diminishing returns to agricultural investment.[5] Ricardo's arguments can be interpreted as a call for the net transfer of a

[5] Of course, Ricardo's argument was not based on the existence of surplus labour, nor on the existence of economies of scale in non-agriculture. Ricardo's argument was essentially based on his theory of comparative advantage and the gains resulting from the removal of tariffs on corn imports.

financial surplus from the agricultural sector through the mechanism of the terms of trade.[6] Similarly, Preobrazhensky's doctrine regarding socialist primitive accumulation in the Soviet industrialization debate was based on this type of net surplus transfer from the agricultural sector – this time to the socialist or the state sector rather than the non-agricultural sector in general (Preobrazhensky, 1980).[7]

Finally, one may also include the dual economy model proposed by Lewis (1954) in the savings-constrained category of models. Lewis's model, with its surplus labour and fixed price assumptions, is a good example of the type of model with persistent disequilibria where industrial accumulation is savings-constrained. The necessary savings for accumulation in the modern sector in Lewis's model, however, is generated within the modern sector itself. The essential assumption in the model is the existence of surplus labour in the traditional sector and the fact that the absorption of this surplus labour in the modern industrial sector does not lead to an increase in the product wage of labour. The assumption of increasing returns in industry is not crucial to the argument, though it could be incorporated. Though Lewis's original model was formulated in terms of a traditional/modern sector distinction, here we shall consider an agricultural/non-agricultural version, with agriculture being the traditional sector and non-agriculture the modern sector. What is essential for Lewis's results in this case is that the shift of surplus labour out of the agricultural sector would automatically increase the productivity of the remaining labour in the sector, and that the increased productivity of labour in agriculture is translated into increased marketed surplus in this sector – in other words, in moving out of the agricultural sector, labour carries its wage basket (equivalent to the wage fund for new employees in industry) with it. Given that, under these circumstances, the output of the industrial sector cannot be entirely exhausted by the demand generated within the sector itself, a portion of the industrial output – equal to the proportion of the wage fund in industry spent on traditional sector wage goods – should be sold to the agricultural sector. As a result, the total sales and purchases of industry from the agricultural sector would

However, aspects of his argument come close to the above interpretation. For example, in *The Principles* (Ricardo, 1951, p. 270) he argues: 'But there is this advantage always resulting from a relatively low price of corn, – that the division of the actual production is more likely to increase the fund for the maintenance of labour, in as much as more will be allotted, under the name of profit, to the productive class, and less under the name of rent, to the unproductive class.'

[6] The different financial mechanisms of surplus flow are explained in Chapter 2.

[7] It should be noted that surplus transfer from agriculture, according to Preobrazhensky's concept of socialist primitive accumulation, served a totally different purpose from all the other ideas discussed here. With Preobrazhensky, this process was meant to lead to a curtailment of the non-socialist sector and its eventual absorption into the state sector – analogous to Marx's notion of primitive accumulation in transition to a capitalist mode of production, whereby the pre-capitalist modes of production are dissolved and their constituent elements are reorganized under capitalist relations. The other theories considered here are more concerned with the complementary interaction between agriculture and non-agriculture in the growth process within a capitalist or mixed market economy framework, and, by and large, abstract from the changes in the social relations of production in the growth process.

balance and there would be no net surplus outflow from the traditional sector. In other words, the entire savings for industrial accumulation is generated within the industrial sector itself.[8] In its original formulation, however, Lewis's model was a general descriptive device rather than a policy-oriented model which could be used directly to analyse the optimal intersectoral resource transfers. Equating Lewis's 'non-commercial' or 'traditional' sector to agriculture, and introducing further assumptions regarding production and technological interdependences between agriculture and industry, the possibility of capital transfers between the sectors, relative rates of return on investment in different sectors, etc., one may use the model to analyse the optimum magnitude and direction of intersectoral resource transfers. It should be added, however, that the level of generality of the Lewis model is such that one may transform it into other types of models (discussed below) by assuming, for example, that the main constraint to industrial accumulation is agricultural marketed surplus (i.e. surplus labour, in moving out of agriculture, does not carry its consumption basket with it, as Lewis implied) or, alternatively, that the binding constraint is industrial demand.

The Marketed Surplus Constraint

The second type of approach in this literature has as its starting-point the assumption that industrial accumulation is constrained by a shortage of foodstuff and agricultural raw materials. In an open economy context, this exemplifies the case of a balance-of-payments-constrained economy, where the growth process encounters inflationary wage spirals due to a shortage of food supplies in the urban areas. Under these circumstances, it is argued that the acceleration of industrial accumulation may necessitate an initial capital transfer to the agricultural sector in order to alleviate the food supply constraint (see e.g. Sen, 1957; Nicholls, 1961, 1963; Dobb, 1964; Kaldor, 1967; Ishikawa, 1967a; Kalecki, 1976, ch. 5). In this approach the relative returns on investment in the agricultural and non-agricultural sectors are immaterial for determining the direction and optimum magnitude of resource flows. Given that the urban food supply is the binding constraint, it can be shown that, even if the rate of return on investment in the industrial sector is higher than in agriculture, accelerating the rate of industrial accumulation in a closed economy would necessitate a transfer of resources to the agricultural sector (assuming fixed industrial real wages). In a neoclassical model, where investment allocation between the sectors takes place through the market adjustment mechanism, such a situation would lead to increased profitability of investment in agriculture through the improvement of agricultural terms of trade. However, given that the marketed surplus constraint is also closely connected to agrarian institutions, and that the lumpiness or riskiness of agricultural investment may inhibit an adequate private sector response, the market adjustment mechanism may lead to an unnecessary

[8] According to Lewis (1954): 'the major source of savings is profit, and if we find that savings are increasing as a proportion of national income, we may take it for granted that this is because the share of profits in the national income is increasing.'

slow-down in the rate of industrial growth. A combination of policies such as taxation of agriculture, direct government investment in agriculture, price policy, institutional change, etc., can produce a more efficient outcome whereby higher growth rates in agricultural marketed surplus can be attained with minimal diversion of investment funds into agriculture.

Once the question of investment allocation between industry and agriculture is explicitly recognized, the formal modelling of the intersectoral relations in the development process becomes extremely complicated. For this reason, all of the descriptive dual economy models which have been developed, either based on Lewis's (1954) classical assumptions or following the neoclassical model of Jorgenson (1961), abstract from the question of agricultural investment.[9] The question of capital allocation between agriculture and industry has, however, been addressed in a number of highly stylized planning models where agricultural marketed surplus is assumed to be the binding constraint. Dixit (1969), for example, sets up a planning model for a food-constrained economy where the problem of minimizing the time required to attain a certain level of capital stock in industry by manipulating the terms of trade and the allocation of investment in the two sectors is investigated. The results signify the importance of investment allocation to agriculture in order to achieve rapid industrialization in a food-constrained economy (see also Dixit, 1973; Hornby, 1968). The final outcome in terms of net agricultural surplus flow depends on assumptions regarding various variables which are normally taken as data in such models, such as the respective rates of return on investment in different sectors, income and price elasticities of demand for food, technologies used in different sectors and their modification, possibilities of effective taxation of agricultural incomes by the government, etc. The direction and optimum magnitude of net intersectoral resource flows in a food-constrained economy thus become empirical questions which depend on the empirically given structure of the economy, the available technologies, the economy's resource endowments and institutions, and the political and economic constraints that they pose for effective government intervention. These considerations shed doubt on whether an aggregate notion of net agricultural surplus can be treated as a policy variable, as seems to have been done in early discussions of industrialization in a dual economy in the savings-constrained approaches discussed above.

The literature on marketed-surplus-constrained industrialization has highlighted two important aspects of intersectoral resource flows in the development process. First is the emphasis on the need for investment in agriculture, and particularly the importance of an inflow of producer goods inputs from industry, which is crucial for sustaining productivity growth in the sector. The second point refers to the relation between this gross inflow of resources and the generation of a net agricultural surplus. This largely depends on the efficiency with which the new

[9] For some useful reviews of these models, see Dixit (1973), Kanbur and McIntosh (1988), and Dutt (1989).

resources are utilized in the agricultural sector and the resulting productivity growth in agricultural production, as well as the institutional arrangements which govern the extent to which the resulting output growth is retained within the sector. This is a theme which will recur time and again in the country case studies in the second part of the book, and is further elaborated in Chapter 3, where we discuss the determinants of net agricultural surplus.

The Demand Constraint

A third possibility is that, rather than being supply-constrained, industrial accumulation might be hindered by lack of investment opportunities arising from the limitation of the home market. This is the case of demand-driven rather than supply-constrained accumulation in the industrial sector which figured prominently in the early debates on balanced growth and the 'big push' theory in the development literature (see Rosenstein-Rodan, 1943; Nurkse, 1953; Hirschman, 1958; Streeten, 1959). The central demand problem in the 'big push' theory, as pointed out in the previous section, resulted from the existence of economies of scale in industry. In the context of dual economy models, however, the demand problem is said to arise from lack of investment incentive which may be due to the sluggish growth of the agricultural sector. A new generation of demand-constrained dual economy models have appeared which, unlike Lewis (1954) or Jorgenson (1962), where industrial profits were assumed to be invested automatically, introduce an independent industrial investment function (see e.g. Bell, 1979; Lysy, 1980; Taylor, 1983). None of these models introduce agricultural output as a direct determinant of industrial investment. But considering that, in a predominantly agrarian economy, agricultural output forms an important exogenous source of demand for industrial products, and taking into account the propagating multiplier accelerator effects, one can easily conceive of a situation in which the pace of industrial investment is set by the growth of demand emanating from the agricultural sector. To ensure that such a model does not hit the food supply constraint before the agricultural demand constraint, however, one may need to assume a generally low investment propensity in the economy (see e.g. Rakshit, 1982).

The demand constraint thesis has recently been put forward in the context of sluggish industrial investment and excess capacity in the Indian industrial sector since the mid-1960s by Bagchi (1975), Raj (1976), Rangarajan (1982), and Mundle (1981, 1985). According to Mundle (1985), this phenomenon is largely to be explained by the over-extraction of agricultural surplus which, on the one hand, has hindered commercialization of agriculture and growth of demand for intermediate goods from industry, and, on the other hand, has led to sluggish growth of final demand for manufactured consumer goods due to low growth of agricultural incomes. As in the case of the marketed-surplus constraint, the policy implications of the demand constraint case also seem inevitably to have involved the issue of the net intersectoral resource flows or net surplus flows between the sectors – as most of the policy instruments such as government taxation and subsidies, price policies, direct investment, etc., are also important components of the net surplus flow between sectors.

On the basis of these arguments, one may be led to conclude that, in a predominantly agrarian economy, where lack of demand is the main constraint to industrial growth, acceleration of industrial accumulation may require a net transfer of investible funds to the agricultural sector. This conclusion, however, is subject to various other preconditions and assumptions, in addition to the basic assumption of a demand-constrained industrial sector. It depends on the combined effect of factors such as the relative responsiveness of agricultural output to investment in that sector, the impact of higher growth of agricultural output on agricultural incomes, the income elasticity of demand for manufactured consumer goods by agricultural households, and the strength of the backward linkages between agriculture and industry. It should be noted further that, in the present-day developing economies, a major proportion of demand for industrial products comes from the growth of government investment and services in the urban areas, and, as pointed out by Kuznets (1964), the significance of agricultural demand for industrial output is expected to decline over the process of industrialization. The mere fact that industrial growth may be constrained by the limitations of the home market, therefore, does not necessarily imply that the optimum direction of intersectoral resource transfer should be towards agriculture, or even that a higher gross inflow of resources into the sector may be called for.

A related, and perhaps more important, argument on the demand side can be put forward in relation to the intersectoral resource flows and their impact on the structure of the home market rather than aggregate demand. According to this view, the method and direction of intersectoral resource extraction may lead to a lopsided structure of domestic demand, and hence industrial output, through its income distributional and employment effects. This may result in what Kalecki (1970) has referred to as a process of 'perverse growth', geared to the luxury consumption of the rich, which may endanger the long-term viability of the growth process. Various formal dual economy models have been constructed along these lines, in which income distribution affects industrial growth through its demand-side influences (see e.g. Taylor and Bacha, 1976; Dutt, 1984; Taylor, 1985). However, to link the income distribution argument to the patterns of intersectoral resource transfers, one requires a disaggregated treatment of the notion of surplus transfer and an analysis of the impact of different mechanisms of resource transfer on income distribution and employment, which will be the subject of later chapters (see particularly Chapter 10).

The Labour Constraint

Finally, there is the case of a labour-constrained economy. In the development literature, the labour constraint facing the developing economies mainly refers to a shortage of skilled labour and managerial know-how. Such economies are usually portrayed as surplus labour economies, where the main problem is to find productive employment for the pool of unskilled labour rather than facing a general labour constraint as such. Consequently, the case of a labour-constrained economy has figured relatively little in the debate on intersectoral resource transfer and economic development. This is not to deny the important role of the agricultural sector in the provision of labour to other sectors of the economy, but rather that, from a policy point of view, a general labour shortage does not seem to be the relevant constraint for these economies. On the contrary, the policy issue is the impact of different intersectoral resource flow mechanisms on employment generation rather than the alleviation of labour shortages.

Though a general labour shortage may not be relevant as a constraint on industrial growth in a dual economy framework, the issue of labour transfer still plays a significant role in the intersectoral resource flow processes. This role partly depends on which of the other constraints discussed above happens to be binding, and partly on the prevailing institutions and the policies adopted by the government. The absorption of surplus agricultural labour in other sectors of the economy would reduce population pressure on land and would be likely to have an immediate effect in the form of increased productivity of labour in the agricultural sector. In a food-constrained economy it may result in higher availability of agricultural products to non-agricultural sectors, depending on the institutional factors which govern the control and disposal of the increased per capita output in the agricultural sector. Similar institutional factors determine the extent to which the increased agricultural labour productivity would translate into higher financial surpluses in agriculture which would aid industrial growth in a savings-constrained economy. In a demand-constrained economy, such labour productivity growth in agriculture would inevitably lead to higher demand for industrial products in the agricultural sector. These effects, of course, would be magnified in the long run if labour productivity growth were further enhanced by technological innovations in agricultural production. This introduces a second important aspect of intersectoral labour transfer, that is, the transfer of skills and know-how in general which normally accompanies the introduction of new inputs and new methods of agricultural production and takes the form of a flow from other sectors into agriculture. This is an aspect of dynamic interaction between different sectors in the intersectoral resource flow process, which is the subject of the next section.

1.3 Dynamic Complementarity and Intersectoral Resource Flows

The assumption of surplus labour has been central to the dual economy models which have been set up to analyse the pattern of intersectoral resource flows in economic development. This assumption is an important precondition for rapid industrialization without necessarily curtailing the growth of the agricultural sector; in other words, accelerating the growth of industrial accumulation need not be at the expense of agricultural growth, as in the neoclassical competitive general equilibrium models. The historical experience of growth and structural change in numerous countries, however, seems to suggest an even stronger complementary relation between agriculture and industry in the growth process, whereby growth of industrial output and productivity is positively associated with that of agriculture.[10] Such dynamic complementarities, though by and large absent from the formal dual economy models referred to in the previous section, introduce an important new dimension to the intersectoral resource flow process in economic development. As pointed out above, in all disaggregated treatments of economic growth in which industrialization is posed as the central problem of economic development such dynamic complementarities are implicitly assumed. In this section we shall take a closer look at the sources of such positive externalities between agriculture and industry, and consider their likely implications for the intersectoral resource flow processes.

Some of the major sources of dynamic complementarity between agriculture and industry in the growth process appear to be technological; that is, they either arise from the particular characteristics of technologies of production in different sectors or they are generated by technological interdependences between the sectors. Static and dynamic economies of scale associated with the industrial technology are examples of the former which, as mentioned in the previous section, played a key role in development theory from its inception in the post-war period, and have come to play an important part in the recent developments in growth theory.[11] In the development literature the main focus of analysis has been on the industrial policy implications of the existence of economies of scale. These are treated as instances of market failure where government intervention is needed to orchestrate industrial investment. In a dual economy framework, however, econ-

[10] For empirical results in relation to a cross-section of advanced and developing economies, see Hwa (1989) and Nishimizu and Page, jun. (1989).

[11] Analysis in the early development literature was focused on static economies of scale internal to the firm and associated with indivisibilities, as highlighted in the famous shoe factory example of Rosenstein-Rodan (1943). A more recent formalization of the 'big push' theory which assumes static economies of scale is provided in Murphy *et al.* (1989). Dynamic economies of scale belong to a macroeconomic concept referring to the scale economies which are external to the firm or even to the industry and which arise from spillovers in learning or economies of specialization (Young, 1928; Kaldor, 1967; Arrow, 1962). Recent endogenous growth theories (see e.g. Romer, 1986, 1990; Lucas, jun., 1988) utilize aspects of this second type of external economies in order to demonstrate how long-term growth rates of countries with access to similar technologies may differ in a conventional neoclassical model.

omies of scale in industry can also form one aspect of the dynamic complementarity between agriculture and industry. The growth of productivity and output in agriculture under these circumstances leads to increased output and productivity in industry by increasing the demand for the industrial sector.

This process, however, is not reciprocal, as agricultural output is normally supply-constrained, and the sector does not exhibit economies of scale.[12] Reciprocity of the dynamic complementarities between agriculture and industry is brought about by technological spillovers of manufacturing growth acting through the supply side of agricultural production. In addition to scale economies, the manufacturing sector is believed to be endowed with special attributes for the extension of capital and machinery, the application of science and technology, and the introduction of new methods of work (Rosenberg, 1976). Furthermore, at low to middle stages of economic development, the income elasticity of demand for manufacturing output tends to be greater than unity. With the combination of these features, the manufacturing sector is capable of achieving rates of growth well above the other sectors of the economy, and the growth of output in that sector can also go hand in hand with the simultaneous growth of employment and productivity of labour. Manufacturing growth is therefore believed to contribute to the growth of the agricultural sector in two ways. On the one hand, its absorption of surplus labour could help to restructure the social and technical conditions of production in the agricultural sector and create the conditions for accumulation of capital and introduction of modern technology (Kaldor 1972*a*, 1972*b*). On the other hand, by providing modern manufactured inputs and know-how, it could make such technical change possible. This is of particular significance with regard to the agricultural sector, where imported technology has to be adapted according to the local ecological relationships and resource availabilities (see Rosenberg, 1976, ch. 9.4; Timmer, 1988).

The growth of the manufacturing sector therefore seems to assume a central role in the development process, as it leads to productivity growth in the sector itself and produces some of the essential preconditions for productivity growth in other sectors of the economy. The strategic role of the manufacturing sector as the 'engine of growth' may necessitate government intervention to direct resources into the sector, at a rate well above that indicated by the market rate of return on investment in different sectors. It does not follow from this, however, that the acceleration of industrial investment would necessarily involve a net surplus outflow from agriculture, as sometimes suggested by the savings-constrained dual economy models. As noted above, an important aspect of the dynamic complementarities between industry and agriculture works through the contribution of industry to productivity growth in agriculture from the supply side. This involves a considerable inflow of new inputs and investment goods from industry into

[12] Note that economies of scale which may exist in agriculture are associated with the size of the farm rather than the size of the market for agricultural products as such.

agriculture. The final outcome, in terms of the net flow of resources, depends in the long run on the efficiency of resource use in agriculture, which in turn depends on agrarian institutions and the existence of appropriate infrastructural conditions for the adoption and efficient use of new technologies in agriculture. In this sense, industrial growth could be said to be a necessary, but not a sufficient, precondition for the growth of agricultural productivity. There may be other reasons why industrial growth in itself may not be sufficient to generate dynamic technological effects in agricultural production. For example, given the highly capital-intensive nature of modern manufacturing technology, growth of manufacturing may not lead to the absorption of surplus labour and the reduction of population pressure on land. The type of industries developed, also, may not contribute to the assimilation and faster diffusion of modern technology in agriculture. Under these circumstances, attempts to increase the rate of industrial investment, no matter how the investment is financed, can easily be aborted by agriculture posing a growing financial burden on the rest of the economy. For example, government attempts to accelerate industrial investment through surplus extraction (e.g. by taxation) from an organizationally and technologically stagnant agricultural sector may be neutralized by adverse internal terms of trade movement and inflationary pressures resulting from the inelastic demand for food. The possibility of such taxation in the first place would also depend on the organizational forms in agriculture, the degree of its commercialization, etc.

On the other hand, in a technologically progressive agricultural sector where new investment and increased input use is combined with fast rates of productivity growth, the complementary relationship between agricultural and industrial growth is most pronounced. Under these circumstances, the increased inflow of inputs into the agricultural sector would give rise to high rates of growth of output and marketed surplus, from which even an increase in the net surplus outflow from agriculture may result. Such an increase need not necessarily be enforced through government taxation or other non-market-mediated means; with an adequate level of development of the financial markets in the countryside, it can take place through voluntary savings and financial asset acquisition by the farm households. This points to an important aspect of the intersectoral resource flows which, by and large, is neglected in the existing dual economy models reviewed in the previous section. The main aspect of intersectoral resource flows considered there seems to be related to the causes and consequences of different patterns of allocation of resources *between* different sectors, with the efficiency and productivity of resource use *within* the sectors assumed as given. The relative significance of these two interrelated aspects of intersectoral resource flows in the development process – namely, the efficiency of resource allocation between sectors and the productivity of input use within the sectors – is, of course, a matter for empirical investigation. Anticipating some of the results of the country case studies in later chapters, it appears that the efficiency of input use and productivity growth in agriculture and industry play a much more important role than dual economy models of intersectoral resource flows have allowed for.[13] In general, it appears that the magnitude and direction of resource flows, and the possibility of government intervention to influence these flows, depend not only on considerations such as the assumed binding constraints, rates of return on investment, and saving propensities, but also on the changes in these variables, which depend on institutional and technological dynamism within the two sectors. Before giving a more detailed consideration to the determinants of intersectoral resource transfer and possible policy choices, however, it is necessary to define more precisely the various notions of agricultural surplus and the different possible mechanisms of surplus transfer. This will be the task of the next chapter.

1.4 Concluding Remarks

In this chapter we have examined different theoretical perspectives on the role of intersectoral resource flows in the development process. The main focus was on the role of agricultural surplus in financing industrial accumulation in transition from a predominantly agrarian economy to a mature economy with a diversified industrial structure. From the policy point of view, the question has normally been posed in terms of the possibility of accelerating the transition process by the adoption of policies which intensify the process of surplus transfer from agriculture for the purpose of industrial capital formation. This question can be considered in two steps: first, the possibility of increasing savings of agricultural households under a given production potential in the sector; and, second, the transfer of the surplus, which is basically one of allocation of investment between the different sectors. Of these two questions, the latter, i.e. the allocation of a given amount of investible surplus between the sectors, seems to have been more prominent in the discussion of intersectoral resource flows in the recent literature. We distinguished between two types of models in the discussion of intersectoral resource flows. On the one hand, there are the full employment competitive equilibrium models of neoclassical economics, according to which attempts to reallocate investment between the sectors would be inefficient and could lead to the contraction of agricultural output and the overall national product. On the other hand, conventional development economics, with the assumptions of surplus labour in agriculture and economies of scale in the industrial sector, has attempted to establish grounds for policy intervention to increase the rate of industrial investment above that warranted by the market-determined rates of return on investment in the different sectors. Despite the similarities in terms of these

[13] The significance of technological progress in economic development has been highlighted in aggregate growth accounting exercises for both the industrialized and developing countries (see e.g. Solow, 1957; Denison, 1962, 1967; Kendrick, 1973; Chenery, 1986; Hwa, 1989). The relationship between technological progress and intersectoral resource flows, however, has generally been neglected in the dual economy models.

fundamental assumptions, conventional development economics has expressed a variety of views with regard to the direction of agricultural surplus flow.

We distinguished between three different types of approach to the question of agricultural surplus transfer in conventional development economics on the basis of what each assumed to be the main constraint on industrial accumulation: namely, the savings constraint, the food constraint, and the demand constraint models. Of these, only the savings-constrained models envisage the need for a net surplus transfer from agriculture in order to accelerate non-agricultural investment in the early stages of industrialization. Under these circumstances, policies which encourage a higher rate of allocation of investment to industry would shorten the transition period in a predominantly agrarian economy, particularly when the mobility of the factors of production is seen to be restricted by the underdeveloped institutions of agriculture, or when the economy is faced with major instances of market failure, e.g. as a result of economies of scale and learning in industrial production. Under the exclusive assumption of savings being the binding constraint on industrial accumulation, the danger of the economy hitting the food or the demand constraint under such a policy regime is also assumed away. In a food-constrained or demand-constrained economy, however, accelerating industrial growth may necessitate an initial transfer of investment funds in favour of agriculture.

Despite the differences in approach and basic assumptions of conventional development economics and the neoclassical theories, they share one key question in their analysis of the intersectoral resource flows in the process of development: namely, they pose the problem as one of the allocation of a *given* amount of investment funds *between* the sectors. While neoclassical economics concentrates on the question of static efficiency of resource allocation between different activities under full employment conditions, the development approach introduces in addition dynamic economies specific to manufacturing industry which can lead to a divergence between private and social rates of return in investment allocation between the different sectors. As we have noted, however, a key variable in the determination of net intersectoral resource flows from a long-term perspective is likely to be the efficiency of resource use *within* the sectors, and, in particular, that of the agricultural sector, which is closely related to technological spillover effects of different patterns of sectoral growth as well as the production institutions of agriculture. In addition, agrarian institutions and the production relations in agriculture also influence the generation of agricultural savings by determining the control and utilization of agricultural surplus by different social classes. This latter point, which has been more prominent in the analysis of sectoral transformation by the classical political economists, gives an additional significance to the structure of, and the change in, agrarian institutions as a determinant of agricultural surplus transfer. For a more detailed study of these points, however, we require a more precise definition of the net agricultural surplus and its constituent elements. This forms the topic of the next chapter.

2

Definitions and Mechanisms of Surplus Transfer

2.1 Introduction

The measurement of intersectoral resource flows involves a multitude of conceptual and data problems which are not often adequately addressed in the literature. The basic problem is the lack of a systematic accounting framework to establish a unified basis for measurement and comparison of the results from different country studies. This has also created a certain degree of ambiguity in the definition and measurement of intersectoral resource flows. The aim of this chapter is to set up an accounting framework which could work both as an aid to conceptual clarity and in facilitating the measurement of intersectoral resource flows. This is done by defining resource flows within a social accounting matrix (SAM) framework.[1]

We start by a brief discussion of SAM as an accounting framework, and proceed to set up a simple SAM with the minimum entries necessary to allow the definition of different notions of agricultural surplus discussed in the literature. Section 2.3 discusses some of the conceptual issues arising in the measurement of intersectoral resource flows, related to the distinction between institutions and sectors and the definition of sector boundaries. Section 2.4 uses the SAM to derive various concepts of agricultural surplus, both from the financial and the real sides. In Section 2.5 we distinguish between intersectoral resource flows in current values and in real terms, and derive a general formula for the terms of trade effect. A summary and the conclusions are presented in Section 2.6.

2.2 A Social Accounting Matrix Framework

In this book we are interested in the use of SAM as a statistical information and accounting system. SAM methodology consists of a matrix representation of the circular flow of income and expenditure as in national accounts, combined with the

[1] For introductory discussions of social accounting matrices, see Pyatt and Thorbecke (1976), Pyatt and Roe (1977), and King (1981). For earlier discussions of the use of a social accounting matrix framework to define intersectoral resource flows, see Karshenas (1989) and Morrisson and Thorbecke (1990).

input–output model of the production sector. Table 2.1 shows a consolidated SAM in which each entry represents a matrix, with the definitions given below the table. As usual, columns represent expenditures or monetary outgoings and rows show receipts or monetary incomings. Sectors or activities produce goods and services by using intermediate products (A) and factor services (F) provided by institutions. Factor incomes received by institutions (Y) are spent on current consumption (C), invested in physical assets (I), or saved (S). The table also shows current transfers between institutions (T), e.g. government taxes and subsidies, migrant workers' remittances, etc., and capital transfers (K), e.g. changes in the financial assets of institutions. Accounting consistency requires that the sum of the values in each row should equal the sum of the values in the corresponding column. For example, the equality of the sums in row 1 and column 1 implies that total sales of each sector should be equal to its gross output; and, in the case of row 3 and column 3, that the income of each institution should be equal to its current expenditure (consumption) plus savings, net of taxes and subsidies and other current transfers. The equality of the sums of row 4 and column 4 means that savings must equal investment.

We have excluded 'the rest of the world' from Table 2.1 and from the subsequent discussion, as it does not help the exposition of different definitions of agricultural surplus discussed in the literature, and it unnecessarily prolongs the formulations. Compared to the national accounts, the only additional information in the table is the inclusion of the input–output table (A), together with the capital account transactions between institutions (K) and some items of current transfers (T) which are not normally included in the national accounts. These additions are essential for the definition and measurement of intersectoral resource flows.

Table 2.1 A schematic representation of social accounting matrix

	Activities	Factors	Institutions (current A/C)	Institutions (capital A/C)
Activities	A		C	I
Factors	F			
Institutions (current A/C)		Y	T	
Institutions (capital A/C)			S	K

A = matrix of intermediate demands of activities.
C = consumption, i.e. current expenditure by institutions on output of activities.
I = investment, i.e. capital expenditure by institutions on output of activities, including stock-building.
F = factor income payments made by activities.
Y = distribution of total factor income among institutions.
T = current transfers between institutions including direct and indirect taxes levied on institutions.
S = saving of institutions.
K = capital transfers between institutions.

In order to discuss the issue of surplus transfer or intersectoral resource flows, we need a more disaggregated SAM with explicit representation of the sectors and institutions of interest. Such a version of the SAM, with minimum adequate entries to represent the intersectoral resource flows between agriculture and the rest of the economy, is shown in Table 2.2. The table distinguishes between two activities or sectors (agriculture and non-agriculture) and three institutions (farm households, government, and others). For ease of presentation, factors of production have been consolidated into one category. Transactions or transfers within sectors and institutions are consolidated out, because they do not enter into intersectoral resource transfers. The rest of the world could be introduced into the table, and factors of production could be further disaggregated. The same applies to taxes on inputs and on capital. However, such modifications would substantially increase the size of the table without adding more insight.

The accounting framework presented in Table 2.2, therefore, is a modified SAM, tailored to address the issue of surplus transfers between agriculture and the rest of the economy in a simple manner. There are, of course, further modifications which need to be made to particular items in the table in order to arrive at a meaningful description of intersectoral resource flows in practice. These will be discussed as we proceed. At this stage, however, the meaning of each row and column corresponds to conventional SAM definitions, as follows:

Row 1: Agriculture as a production activity receives money from sale of its products as intermediate goods to other sectors (A_{an}), and as consumer goods (C_{af}, C_{ag}, and C_{ao}), and investment goods (I_{af}, I_{ag}, and I_{ao}) to institutions. The sum of these proceeds constitutes the gross output or gross sales of the agricultural sector.

Column 1: The proceeds from gross sales of the agricultural sector is partly spent on purchases of intermediate products from the non-agricultural sector (A_{na}), and the remaining part, which is the value added in the agricultural sector, is paid out to the factors of production (F_a).

Row 2: The gross output of the consolidated non-agricultural sector is composed of sales of intermediate products to the agricultural sector (A_{na}) plus sales of final products to institutions for consumption (C_{nf}, C_{ng}, and C_{no}) and investment (I_{nf}, I_{ng}, and I_{no}).

Column 2: The proceeds from gross sales of the non-agricultural sector is partly spent on purchasing intermediate products from the agricultural sector (A_{an}), and the rest, which goes to the remuneration of the factors of production, is the value added in the sector (F_n).

Row 3: Factors of production receive incomes from the agricultural sector (F_a) and the non-agricultural sector (F_n). Further disaggregation of this row would have shown the social form that these factor incomes take (e.g. wages, profits, and rent). However, since our present purpose is to analyse the movement of factor

Table 2.2. A social accounting framework for intersectoral resource flows

	Activities		Factors	Institutions (current A/C)			Institutions (capital A/C)		
	Agriculture	Non-agriculture		Farm hhlds	Government	Others	Farm hhlds	Government	Others
Activities									
1. Agriculture		A_{an}		C_{af}	C_{ag}	C_{ao}	I_{af}	I_{ag}	I_{ao}
2. Non-agriculture	A_{na}			C_{nf}	C_{ng}	C_{no}	I_{nf}	I_{ng}	I_{no}
3. *Factors*	F_a	F_n							
Institutions (current A/C)									
4. Farm households			Y_f		T_{fg}	T_{fo}			
5. Government			Y_g	T_{gf}		T_{go}			
6. Others			Y_o	T_{of}	T_{og}				
Institutions (capital A/C)									
7. Farm households				S_f				K_{fg}	K_{fo}
8. Government					S_g		K_{gf}		K_{go}
9. Others						S_o	K_{og}		

incomes between sectors and institutions rather than their social form, it is more convenient to treat them in aggregate.

Column 3: The factor incomes are distributed amongst institutions which provide factor services. For example, farm households receive the value added from their farming activities net of rents and interest payments to non-farming landlords and moneylenders (part of Y_f). The government receives profits from the state-owned enterprises (part of Y_g); urban non-agricultural workers receive wages (part of Y_o), etc.

Row 4: Farm households receive primary income (Y_f) in return for the factor services that they provide, as discussed above. In addition, they receive direct transfers from the government (T_{fg}) and from the other non-governmental private institutions (T_{fo}). The total of this row constitutes the personal (pre-tax) income of the farm households.

Column 4: Income received by farm households is either consumed (C_{af}, C_{nf}), paid out as direct taxes to the government (T_{gf}), transferred to other non-governmental institutions (T_{of}), or saved (S_f).

Row 5: Government income consists of primary income (Y_g) plus direct taxes on farm households (T_{gf}) and other institutions (T_{go}). Indirect taxes on agricultural and non-agricultural products could have been included here also, at the intersection of this row and columns 1 and 2 respectively.

Column 5: Government revenue is disposed of in the form of current government consumption expenditure (C_{ag}, C_{ng}), direct subsidies to the farm households (T_{fg}) and the other institutions (T_{og}); the residual is saved (S_g).

Row 6: Other institutions receive primary income in the form of rents, profits, and wages in return for their factor services (Y_o). This, together with transfers from the farm households (T_{of}) and direct subsidies from the government (T_{og}), adds up to their total income.

Column 6: Total income of other institutions is spent on consumption (C_{ao}, C_{no}), paid out in current transfers to the farm sector (T_{fo}) or as direct taxes to the government (T_{go}), with the remainder being saved (S_o).

Row 7: The farm households acquire capital funds through their own savings (S_f) and capital transfers from the government (K_{fg}) and other private financial institutions (K_{fo}).

Column 7: The capital funds acquired by the farm households are invested either in the form of physical assets (I_{af}, I_{nf}) or in acquiring financial assets from, or lending to, the government (K_{gf}) or the private sector (K_{of}).

Row 8: Similarly, the government acquires capital funds either through its own savings (S_g) or capital transfers from the farm households (K_{gf}) or the other institutions (K_{go}).

Column 8: The government's capital spending takes the form of investment in physical assets (I_{ag}, I_{ng}) or the acquisition of financial assets from (lending to) the farm households (K_{fg}) or other institutions (K_{og}).

Row 9: Total capital funds at the disposal of other institutions are composed of their own savings (S_o) plus the capital funds acquired from the farm households (K_{of}) and the government (K_{og}).

Column 9: Other institutions dispose of their capital funds by investing in physical assets (I_{ao}, I_{no}) or by acquiring financial assets from the farm households (K_{fo}) or the government (K_{go}).

2.3 Some Conceptual and Measurement Problems

A first glance at Table 2.2 would at once identify an important conceptual problem in defining the intersectoral resource transfers: namely, that consumption, investment, and current and capital transfers take place within and between institutions, while production takes place in sectors. As a consequence, even in the simplified case being considered here, resource flows seem to exhibit diverse and incompatible origins and destinations. For example, intermediate inputs form a flow between production sectors, while consumption and investment constitute flows between production sectors and institutions, and capital and current transfers represent flows between institutions. This implies that intersectoral resource flow only becomes a meaningful concept once we have redefined sectors and institutions in such a way as to create a correspondence between the sectors and institutions of interest. As Ishikawa (1967*a*) has pointed out, aggregate surplus transfer is only meaningful in the context of institutions, and production sectors should be chosen so that they incorporate the activities of the respective institutions. For example, if the resource transfer between the farm sector and non-farm sector is of interest (as in Table 2.2), then the activity referred to as 'agriculture' in the table should incorporate all the production activities of the farm sector, including any non-agricultural activities.

Failing this, our resource flow measure would be a hybrid measure where, for example, the flow of intermediate products refers to agricultural/non-agricultural activities, and the consumption flow refers to farm/non-farm households. To correct for this requires a certain degree of rearrangement of categories as defined in conventional SAM and national accounts – depending, of course, on the availability of the necessary data and information. Unfortunately, this point has not been adequately addressed in the empirical literature, probably for lack of data. Often it is not possible to find accurate data on non-agricultural activities of the farm sector. However, it may be possible to assess the magnitude of error by examining the different sources of income of agricultural households from the household budget surveys.

A second conceptual issue is the appropriate choice of institutional or sector boundaries. In addition to the conventional broad economic sectors, the literature suggests various other concepts such as farm/non-farm, subsistence/commercial, rural/urban, organized/unorganized, etc. The choice obviously depends on the purpose of the study and its theoretical starting-point, as well as the availability of data. The rural/urban distinction has been suggested in studies which concentrate on the political dimensions of resource allocation. For example, Lipton (1977) invokes this categorization in arguing for the existence of strong urban bias in resource allocation in developing countries, which is said to have arisen because of political power commanded by town dwellers.[2] The rural/urban distinction side-steps the sectoral boundaries which have been the central focus of economic analysis of resource flows since the time of Quesnay (1758), and none of the empirical case studies considered in this book is based on such a classification. We shall therefore leave out this distinction in the subsequent discussion, though the empirical results presented here may be of interest to the urban bias debate as well.

Sectoral distinctions based on the organizational form of production, such as formal/informal, subsistence/commercial, capitalist/non-capitalist, etc., introduce an important social dimension to the analysis of intersectoral resource flows. As we noted in Chapter 1, some of the influential two-sector models in the economic development literature (e.g. Lewis, 1954) are based on such a classification. The distinction of sectors on the basis of organizational forms of production was also central to the economic analysis of the classical political economists, and particularly to Marx's analysis of transition to the capitalist mode of production (see e.g. Marx, 1976, chs. 26–33). This approach highlights the significance of the particular social form that the surplus takes, and to some extent the decomposition of surplus transfer from the financial side (discussed below) comes close in providing the empirical categories necessary for this type of analysis. However, to combine an analysis of resource flows between different organizational forms of production with that of material and technological interdependence between conventionally defined sectors (e.g. agriculture/industry) requires a multisectoral model in which each sector is subdivided into different organizational forms.[3] Such an exercise would lead us far afield, and its empirical requirements would be far beyond the capacity of the available data. This is not, however, to neglect the significance of the social form that intersectoral resource flows take under different organizational set-ups. Indeed, a main conclusion of this book (which we shall elaborate in later chapters) is the key role that the organizational forms of production play in the efficiency of resource use within each sector and, indeed, in determining the magnitude and direction of resource flows.

Both analytical simplicity and the nature of the available data require that

[2] According to Lipton (1977, p. 269): 'In a less-developed country "the executive of the modern state is but a committee for managing the common affairs" not of the bourgeoisie but of the townsmen.' For a critique of the urban bias thesis, see Byres (1979).

[3] For the discussion of a model along these lines in relation to the Indian economy, see Bharadwaj (1979).

sectoral boundaries are defined here in relation to the type of product, along the lines of conventional industrial classification discussed, for example, in United Nations (1968). More specifically, we follow the relatively aggregative agriculture/non-agriculture distinction. The aggregation of all the non-agricultural activities into one sector is a simplification which is permitted by the fact that we are primarily interested in the issue of resource transfers from the point of view of the agricultural sector. As was pointed out above, however, the agricultural sector should be redefined to include all the processing activities within the farm households not conventionally classified as agriculture.

There are a number of other modifications which one needs to make to the classifications of the SAM described in the previous section in order to derive accurate formulae for the intersectoral resource flows. These are discussed in the context of the actual derivation of these formulae in the next section.

2.4 The Real and Financial Mechanisms of Resource Transfer

Having defined clear sector boundaries and the corresponding institutional classifications, we may proceed to the analysis of different notions of agricultural surplus and the mechanisms of surplus transfer. This can be done in a straightforward manner by utilizing the accounting identities implicit in the SAM presented in Table 2.2. In what follows, the institutional boundaries are defined in terms of the farm/non-farm dichotomy (as in Table 2.2), and the agricultural sector is redefined correspondingly. We shall therefore use agriculture/non-agriculture and farm/non-farm surplus transfers interchangeably.

An important concept of surplus transfer in the literature is the 'net finance contribution' of the agricultural sector to accumulation in other sectors of the economy. As we have seen in the previous chapter, this concept is significant in situations in which the savings constraint is believed to be the main bottleneck to industrial accumulation. On the real side, this is defined as the difference between commodity exports and imports of the agricultural or the farm sector to the rest of the economy. Millar (1970) has thus called it the 'net product contribution' of the agricultural sector. If we denote agricultural exports or sales by X_a and imports or purchases by Ma, the net finance contribution of agriculture is $R = X_a - M_a$. From rows 1 and 2 of Table 2.2 it can be seen that X_a and M_a can be decomposed according to the type of demand in the following way:

$$X_a = A_{an} + C_{ag} + C_{ao} + I_{ag} + I_{ao}; \tag{2.1}$$

$$M_a = A_{na} + C_{nf} + I_{nf}. \tag{2.2}$$

For this decomposition to be precise, a further modification of the conventional interpretation of the SAM classifications is necessary. It requires the institutional demand (particularly for investment goods) to be identified also as the sector of destination. For example, I_{nf}, or investment in the farming or agricultural sector composed of goods purchased from the non-farm sector, should include all such investments in the farm sector, whether financed by the farm households or by other institutions, e.g. the government. This involves a simple reclassification of the SAM entries which requires no additional data or information apart from that which has already been used in constructing the SAM. Simply, government investment in agriculture will be incorporated into I_{nf} and I_{af}, with a corresponding addition to K_{fg} in the form of capital transfers from the government to the farm sector.[4] With this modification, the interpretation of Equations 2.1 and 2.2 is straightforward. Equation 2.1 tells us that the sales of the agricultural sector to the non-agricultural sector is composed of sales for intermediate use (A_{an}) plus those for consumption use (C_{ag}, C_{ao}) and investment use (I_{ag}, I_{ao}) in the government and other institutions. Equation 2.2 decomposes the purchases of the agricultural sector into those for intermediate use (A_{na}), consumption (C_{nf}), and investment (I_{nf}).

The above formulae are often used to measure the intersectoral resource flows from the real or commodity flow side. They clearly show the source of error due to misclassification of sectors and institutions referred to in the previous section. While consumption categories refer to households as institutions, the intermediate and investment demands are normally measured on the basis of sectoral or activity demands. Thus, it is important to ensure that there is a correspondence between the sectors and institutions of interest, or at least to be aware of the magnitude of error involved if the data do not lend themselves to such a reclassification.

Given the accounting identity between column 1 and row 1 in Table 2.2, Equation 2.1 can also be written as:

$$X_a = A_{na} + F_a - C_{af} - I_{af}. \tag{2.3}$$

Combining Equations 2.2 and 2.3, we get the net finance contribution of the agricultural sector as:

$$R = X_a - M_a = F_a - (C_{af} + C_{nf}) - (I_{af} + I_{nf}), \tag{2.4}$$

that is, value added in the farm sector minus the total consumption and investment in the sector. It should be noted that this formula is derived from a simplified SAM tailored to contain only some minimum essential features to demonstrate the use of SAM as an accounting framework. Direct taxes, for example, are consolidated into one item, and the rest of the world is ignored. If we were to distinguish between different types of direct taxes in the SAM shown in Table 2.2, then they would be explicitly appearing on the right-hand side of Equation 2.4. Similarly, the inclusion of the rest of the world in the table would have meant the addition of an extra term on the right-hand side for the net foreign exchange contribution of the farm or

[4] A similar operation is needed, of course, for the direct investment of other institutions in the farm sector. This takes place largely when other institutions take on the role of farmers and directly invest funds in the agricultural sector. This is a rather rare event, and, compared to government investment in the agricultural sector, the size of such investment is small and can be ignored in many cases.

agricultural sector to the balance of payments. This latter addition would be useful if one were interested in the foreign exchange contribution (burden) of a particular sector rather than intersectoral resource flows as such. Equation 2.4 as it stands, therefore, is a simple representation of intersectoral resource flows in which the accounts for the rest of the world are consolidated in the income and commodity flow values shown on both sides of the equation. Apart from its simplicity, this equation has the advantage of representing the essential elements in the intersectoral resource flows which we shall use in the following chapter in analysing the determinants of agricultural surplus transfer.

The Financial Mechanisms of Resource Transfer

This concept of intersectoral resource flow can also be looked at from the financial side. A net export surplus of the agricultural sector corresponds to a net outflow of funds from the sector. To derive the financial counterpart of this concept of agricultural surplus, we utilize the accounting identities of column 4 and row 4 and column 7 and row 7 in Table 2.2 (i.e. the current and capital accounts of the farm sector) to get:

$$(C_{af} + C_{nf}) = (Y_f - S_f) + (T_{fg} - T_{gf}) + (T_{fo} - T_{of}); \tag{2.5}$$

$$(I_{af} + I_{nf}) = S_f + (K_{fg} - K_{gf}) + (K_{fo} - K_{of}). \tag{2.6}$$

Substituting these measures of agricultural consumption and investment into Equation 2.4 gives:

$$X_a - M_a = (F_a - Y_f) - (K_{fg} - K_{gf}) - (K_{fo} - K_{of}) - (T_{fg} - T_{gf}) - (T_{fo} - T_{of}). \tag{2.7}$$

The right-hand side of the equation gives the financial counterpart of the surplus transfer. A brief description of each term, in the same order as they appear in Equation 2.7, is as follows:

$(F_a - Y_f)$: the value added in the agricultural or the farm sector minus the net factor income of the farm sector. It thus represents the outflow of net factor incomes from the agricultural or farm sector, mainly comprising rent payments to absentee landlords, interest on loans, and net labour income of the farm sector from non-farm activities (with a minus sign).[5] Had we distinguished between different social forms of factor income in Table 2.2 – namely, wages, profits, interest, and rent – these categories would have been explicitly represented on the right-hand side of the equation. The first term in Equation 2.7, therefore, represents the consolidated net factor outflow from the agricultural sector.

$(K_{fg} - K_{gf})$: the net capital transfers from the government to the farming sector.

[5] Note that the latter category only includes the labour income of the members of the farm household working outside the farm sector. Factor income arising from the non-agricultural activities within the farm household, on the basis of the reclassification discussed in the previous section, is netted out by an equivalent increase in the value added of the farm sector – unless, of course, it is transferred to the non-farm sector through, for example, interest payments to moneylenders.

Since the public sector banks are normally classified along with other banks in 'other institutions', and as the acquisition of government bonds by farm households in developing countries is a rarity, this term is largely composed of net government investment in the agricultural sector.

$(K_{fo} - K_{of})$: the net private capital transfers into the farm sector from the other sectors of the economy. This is mainly comprised of lending to, and borrowing from, formal and informal financial institutions, hoarding and dishoarding of money, acquisition of financial assets, etc. As noted above, lending by public sector banks is also normally classified as a separate item in this category in most of the empirical studies.

$(T_{go} - T_{og})$ and $(T_{of} - T_{fo})$: respectively the net inflow of government and private sector current transfers. The former primarily consists of net direct taxes/subsidies, and the latter largely of remittances of migrant members of the farm household.

Combining the factor payments and current transfers into one term, V, and the capital transfers into K, we get the famous formula:

$$R = X_a - M_a = V + K, \tag{2.8}$$

first discussed by Ishikawa (1967*a*) and often quoted in the later literature. The presentation of the intersectoral resource flows from the financial side, particularly in its decomposed form as in Equation 2.7, brings to light the various mechanisms through which resource transfer from agriculture can take place. Given the variety of mechanisms involved and their differential effect on agricultural development as well as the utilization of the surplus, the usefulness of such a decomposition becomes apparent. Another possible use of financial characterization of the surplus flow is for empirical estimation when data on the real side are incomplete (see e.g. Chapter 7), or for double-checking the accuracy of the real side measures.

Other Notions of Agricultural Surplus

The SAM framework shown in Table 2.2 can be used to derive other notions of agricultural surplus discussed in the literature. For example, marketed surplus is the total sales of the farm sector to the non-farm sector, which is equal to X_a, already defined. Following Millar (1970), we can define the concept of 'net agricultural surplus' as the value added in the farm sector minus the consumption of the farm households. In the notation of Table 2.2 this can be defined as:

$$NS_a = F_a - (C_{af} + C_{nf}) = (X_a - M_a) + (I_{af} + I_{nf}),$$

or

$$NS_a = F_a - C_a = (I_a + X_a - M_a), \tag{2.9}$$

where C_a and I_a represent total consumption and investment of the farm sector respectively. As can be seen, the difference between this notion of agricultural surplus and the net finance contribution of agriculture defined previously is total investment in agriculture. Net agricultural surplus is a useful notion, as it refers to resources made available by the agricultural sector for investment within the sector itself and utilization in other sectors, including exports. Other definitions of agricultural surplus given by Millar (1970) and Ellman (1975) could also easily be derived from Table 2.2, but since these are of limited usefulness for our present purpose we shall not attempt this.

Another concept of agricultural surplus which has been put forward in some of the empirical studies is the concept of savings surplus of the agricultural sector (see e.g. Mundle and Ohkawa, 1979; Mody *et al.*, 1985). The savings surplus of the agricultural sector is defined as the net financial contribution of agriculture as contained in Equation 2.4 above, plus the inflow of net factor income and current transfers into the agricultural sector (which equals V with a minus sign, as defined in Equation 2.8). Adding these two items on both sides of Equation 2.4, we get the decomposition formula for the savings surplus of agriculture as:

$$AS = X_a - M_a - V = [Y_f + (T_{fg} - T_{gf}) + (T_{fo} - T_{of})] - (C_{af} + C_{nf}) - (I_{af} + I_{nf}), \quad (2.10)$$

which, on substitution in Equations 2.7 and 2.8, gives:

$$AS = X_a - M_a - V = K = -(K_{fg} - K_{gf}) - (K_{fo} - K_{of}). \quad (2.11)$$

As can be seen, the savings surplus of the agricultural or the farm sector is not more than a measure of net capital transfer to the other sectors of the economy. This measure is sometimes interpreted as the net contribution of the agricultural or farm sector to investment in other sectors of the economy (see e.g. Mundle and Ohkawa, 1979; Mody *et al.*, 1985; Sheng, 1992). This interpretation, however, holds only for a particular view of what constitutes the contribution of a sector to the growth of the economy.

Note that, in measuring the net finance contribution of agriculture in Equation 2.4, X_a and M_a are defined as the sales and purchases of goods and non-factor services of the agricultural sector to and from the rest of the economy. The new measure, i.e. the savings surplus of agriculture, also includes the sale of factor services and current transfers as contributions of agriculture. For example, the receipts of wage labour and the remittances of members of farm households working in other sectors of the economy are also counted as income which in some sense is contributed by the agricultural sector, and not as an inflow of income originating in other sectors. Similarly, rents to absentee landlords and interest payments on loans from the banks or the middlemen by agricultural households are taken to constitute income contributed by sectors other than agriculture. We do not subscribe to this view here, and rather take the view which is more consonant with that of the classical political economists, that the contribution of a particular sector to the growth of other sectors depends on the productivity of the activities located in that sector rather than the productivity of the factors of production which in some sense are said to belong to that sector. We will therefore be mainly concentrating on the net finance or net product contribution of agriculture rather than the savings surplus as defined above. However, given that net capital transfers are in any case a part of the net finance contribution of agriculture, we shall also refer to this latter measure in the empirical chapters.

2.5 The Terms of Trade Effect

So far we have been considering the intersectoral resource flows at current prices. But, from a developmental point of view, the contribution of agriculture to economic growth depends on the 'real' value of the resources made available. The problem here is basically that surplus transfer consists of the net value transfer of different types of commodities, and therefore one needs a price vector to evaluate the different bundles of goods. The choice of the prices would clearly affect the magnitude and even the direction of net resource flows. The issue, therefore, is to establish what is deemed to be an 'appropriate' price vector to evaluate the composite commodities.

To see the problem more clearly, let us consider it in the context of a simple economy in which agriculture produces a single commodity (say grain) and uses a single input which it purchases from the other sectors (say a specific type of fertilizer), both measured in quantity terms in tons. The exchange value of fertilizer to grain is 1 to 1, that is, if we take the grain to be the numeraire, the price of fertilizer is 1. Let us assume that, in the initial period, 10 tons of grain is exchanged for 10 tons of fertilizer, and therefore the net intersectoral resource flow at current prices is zero. We further assume that the economy is in a stationary state and that net intersectoral resource flows in all periods, at the prevailing rate of exchange of 1, remain zero.

Now let us consider the effect of a change in relative prices or terms of trade between agriculture and industry (grain/fertilizer price ratio) on the intersectoral resource flows. Let us assume that the price of fertilizer in terms of grain increases to 2 in the second year, implying a deterioration in the terms of trade of the agricultural sector. Since we are only interested here in measurement issues we shall continue to assume that everything else remains the same; notably, that the quantities exchanged between agriculture and non-agriculture remain unchanged at 10 tons per year. Despite the constancy in intersectoral exchanges in use value or quantity terms, the value measure of net intersectoral resource flows now depends very much on relative prices. Valued at current prices, the net intersectoral resource flows in the first year remain at zero, but in subsequent years they record an inflow into the agricultural sector equal to 10 units of the numeraire. Valued at the relative prices prevalent in the first year, they retain the value of zero throughout the period, and valued at the relative prices of the second year onwards, they record an inflow into the agricultural sector for all years, including the first year. The results, therefore, very much depend on the valuation or the price system adopted.

It is important at this stage to distinguish between two different sets of issues in the valuation of net resource flows over time. One set of issues relates to the fact that intersectoral resource flows over time have two components, the visible and the invisible. The first, or visible, component is the value of net financial resources transferred from one sector to another, assessed at some base year prices. The invisible component is the terms of trade effect or the change in resource flows resulting from relative price changes. The second set of issues relates to the choice of the base year prices or reference point prices. We shall return to this shortly, but let us begin by considering the first set of issues, assuming that the base year or reference point prices are given.

In the context of the simple example given above, let us assume that the relative prices of the first year are the 'correct' reference point prices. The task is to measure the real value of net resource flows over time in terms of its two constituent elements, namely the visible and the invisible components. This seems all too straightforward for our simple example. We know by assumption that the quantity of commodities exchanged over time has not changed, and, indeed, that the value of net intersectoral commodity exchanges at base year prices remains zero for the whole period. Measuring resource flows from the financial side, however, we see that there has been a net financial flow into the agricultural sector equal to 10 units of the numeraire at current prices. We may then deflate this by agricultural sales or purchase price index, or any other composite price index which we may deem appropriate, to arrive at the real value of the visible component of net resource flows from the financial side. Clearly, this would not be equal to zero, as it does not take into account the invisible component of net resource flows, i.e. the financial loss to the agricultural or the farm sector arising from the adverse movements of the terms of trade. The issues involved here could perhaps be explained better by a more general and formal formulation of resource flow mechanisms.

Denoting real magnitudes by lower-case letters, and using P_x and P_m as the price indexes of the agricultural sector's sales and purchases respectively, the real net product contribution of the agricultural sector (r), valued at base year prices, could be written as:

$$r = (X_a/P_x) - (M_a/P_m) = x_a - m_a. \tag{2.12}$$

This is the real product contribution of the agricultural sector from the point of view of the economy as a whole. What matters here is the net real value of resources made available by a sector or an institution, and the intersectoral relative prices are immaterial in this respect. From a sectoral point of view, however, the real value of the net financial surpluses depends on the terms of trade between the sales of that sector and its purchases. A rise in relative prices in favour of the farm sector, for example, implies an income gain for the sector, at the expense of other sectors. The magnitude of this gain, however, depends on the price index which is assumed to be appropriate for deflating the net financial surpluses/deficits of the sector. To get a clearer idea of this point, let us assume P to be such a price index, enabling the real value of the financial surplus of the agricultural sector, viewed from the point of view of the sector itself (i.e. inclusive of terms of trade gains), to be written as:

$$r' = (X_a/P) - (M_a/P) = (V + K)/P = R/P. \tag{2.13}$$

The difference between r and r′ in Equations 2.12 and 2.13 is the income terms of trade gains of the agricultural sector, denoted by TT. It therefore follows that:

$$r = x_a - m_a = (X_a/P) - (M_a/P) - TT. \tag{2.14}$$

From this and Equation 2.12 we get the general formula for the terms of trade gains for the farm sector as:

$$TT = x_a(P_x/P - 1) + m_a(1 - P_m/P). \tag{2.15}$$

The choice of an appropriate price index (P) has been subject to an old, and as yet, unresolved controversy in the literature (see e.g. Stuvel, 1956; United Nations, 1968; Kurabayashi, 1971; Gutmann, 1981). Ishikawa (1967*a*) has suggested the use of P_x for an export surplus and P_m for an import surplus. Assuming $P = P_x$ in Equation 2.14, we can therefore write the following expression for the real net finance contribution of the agricultural sector:

$$r = x_a - m_a = (V + K)/P_x - TT, \tag{2.16}$$

where $TT = m_a(1 - P_m/P_x)$ is the income terms of trade gains of the farm sector, or the invisible component of agricultural surplus flows, and $(V + K)/P_x$ is the visible component. Thus, in addition to the current and capital transfers discussed in the previous section, the real net finance contribution of the agricultural sector has thus an additional component, that is, the terms of trade effect.

It should be noted that terms of trade improvements in favour of a particular sector do not necessarily imply improvements in profitability. Similarly, the income gains from such relative price changes as those measured above do not necessarily lead to a relative increase in factor remuneration for that sector. It all depends on the relative factor productivity movements in the two sectors, as determined by their respective rates of technological progress. Terms of trade movements are often the result of differential productivity improvements in different sectors, and act as a mechanism for transmitting the benefits of productivity gains in one sector to the rest of the economy through price reductions in the more progressive sector. To measure the effect of terms of trade movements on factor remuneration, one needs to calculate the factoral terms of trade, first discussed by Viner (1937) in the context of the theory of international trade. The factoral terms of trade are calculated by multiplying the commodity terms of trade by an index of relative factor productivities in the two sectors.[6] It may happen that, while commodity terms of trade are deteriorating in a particular sector, the factoral terms of trade turn out to be improving. This may indicate that the gains connected with the improved methods of production in one sector are being transferred to the lagging sector through relative price movements. As we shall see in Chapter 9, this seems to have been an important element in financing agricultural investment in countries that are witnessing a rapid process of industrialization. Of course, how the gains in factoral terms of trade are distributed between different factors of production within each sector depends on the movement of factor prices and factor productivities within the sector – e.g. the movement of wages relative to labour productivity.

In calculating the real intersectoral resource flows so far we have been taking the base year or reference point prices as given. In doing so, we have been treating price indexes such as P_x and P_m as unambiguously defined. This is not realistic, however, and relaxing this assumption reveals that real intersectoral resource flows can be very sensitive to the base year or reference point price system. To illustrate this, let us return to the simple example discussed at the beginning of this section. Assuming the first-year prices to be the reference point prices in that example, we can very easily calculate real net resource transfers from the agricultural sector for the second year onwards from Equation 2.16 as:

$$r = x_a - m_a = (V + K)/P_x - m_a(1 - P_m/P_x),$$

or:

$$r = 10 - 10 = -(10)/1 - 10(1 - 2/1) = 0.$$

The expression on the right-hand side of the second equality sign gives the visible and invisible components of net resource flows in real terms. There is a financial inflow equal to 10 units of the numeraire to the agricultural sector (hence the minus sign), which in this case is neutralized by an equivalent squeeze of resources from the sector through the terms of trade effect. Similar results will be obtained if we take the agricultural purchase price index (P_m) as the deflator for calculating the terms of trade effect. In that case, the formula for real net intersectoral flows becomes:

$$r = x_a - m_a = (V + K)/P_m - x_a(P_x/P_m - 1),$$

which, on substitution of values from our simple example, gives:

$$r = 10 - 10 = -(10)/2 - 10(1/2 - 1) = 0.$$

[6] Viner (1937) distinguishes between the single factoral terms of trade and the double factoral terms of trade. The former is calculated by multiplying the terms of trade of a particular sector by an index of factor productivity in that sector alone. It is a measure of how many units of the goods of the other sector can be obtained for the services of one factor unit of the sector in question. The double factoral terms of trade, which are the ones defined in the text above, measure the number of services (in terms of the other sector's factor units) which can be obtained for one factor unit of the sector in question.

The results would be the same if we took any other composite price index (e.g. a weighted average of P_m and P_x) as appropriate for calculating the terms of trade effect. The choice of the price index would, of course, affect the value of the visible and invisible components, but the value of net intersectoral resource flows would remain the same.

With a change in base year or reference point prices, however, both the value of net intersectoral resource flows and its visible and invisible components can change dramatically. To check this, let us take the second year as the base year. In fact, let us assume that our data start from the second year and that we are unaware of the relative prices that prevailed in the first year. We know that, at current prices, there is a net inflow of resources into the agricultural sector equal to 10 units of the numeraire for all years. However, since there is now no change in prices, this is also a measure of net intersectoral resource flows in real terms, with the terms of trade effect being equal to zero. Using the general formula developed above (assuming P_x to be the price deflator for calculating the terms of trade effect), we can measure the real net resource flows into the farm sector for the third year onwards in terms of the values prevailing in the second year as:

$$r = x_a - m_a = (V + K)/P_x - m_a(1 - P_m/P_x),$$

or

$$r = 10 - 20 = -(10)/1 - 10(1 - 1/1) = -10.$$

The change in the base year, therefore, seems to have altered the results dramatically. There is no terms of trade effect any longer, and the overall measure of intersectoral resource flows has changed from zero to an inflow into agriculture equal to 10 units of the numeraire, which is 100 per cent of the total sales of the agricultural sector. On closer inspection, however, it becomes apparent that the major difference between this case and the one examined above (with the first year as the base year) is that the terms of trade are no longer present. The choice of the base year essentially affects the terms of trade or the invisible component of the intersectoral resource flows.

Of course, relative price changes as dramatic as those above rarely happen in real economies, and the effect of the change in the base year from one year to the next is seldom as pronounced. The problem, however, is an important one, related more to the choice of an appropriate reference price system rather than the choice of a base year as such. The question is really: what relative price system can serve as a better reference point for calculating the terms of trade effect? The answer, in my opinion, very much depends on the purpose of the analysis. Clearly, in the context of the simple example discussed above, the limited amount of information means that it is not possible to arrive at any sensible answer, whatever the purpose of the analysis. This example gave a series of prices and quantities for the commodities exchanged, without any other information about the economy. On this basis, any year could have been as good (or as bad) as any other as a base year. In real economies, however, there is plenty of other information to be drawn upon in selecting the base year prices. It is the way in which this extra information is used to determine the 'appropriate' base year prices that seems to be largely influenced by the aims of the analyst and his/her theoretical starting-point. Of course, the nature of the available information itself also dictates the extent to which one can come close in practice to what is deemed in theory to be an appropriate reference point price system.

It seems plausible as a starting-point to argue that the search for a base year price system is in fact a search for an appropriate set of prices which reflects the opportunity cost to the economy of the exchanged resources in the intersectoral resource flow process. According to economic theory, market prices in an economy in perfect competitive equilibrium reflect the opportunity cost of the resources. On this basis, some economists have searched for an absolute or 'correct' price system which reflects the relative prices which would have prevailed in the economy under perfect competitive equilibrium conditions. One reason often quoted for the divergence between actual prices and competitive equilibrium prices is the 'distortions' introduced by government intervention – particularly those arising from trade and industrialization policies pursued by the governments of developing countries. There are, however, a host of other reasons, perhaps more fundamental than government intervention, for such disparities. Lack of factor mobility and market segmentation, persistence of non-capitalist institutions, particularly in the agricultural sector, and monopoly power by individual producers, particularly in modern industry, are amongst some of the prominent reasons for the divergence between prevailing market prices and competitive equilibrium prices. If we add dynamic considerations such as economies of learning, dynamic economies of scale, and uncertainty combined with irreversibility, then the concept of a competitive equilibrium price system would itself come under question (see e.g. Kaldor, 1978).

Given that the prevailing market prices are unlikely to reflect the opportunity cost of the resources, it becomes imperative to calculate base point prices as shadow prices in an intertemporal optimization exercise. This introduces a certain degree of indeterminacy when choosing base point prices, reminiscent of the one encountered in selecting base year prices in the example discussed above. Shadow prices are likely to vary, depending on the objective function chosen for the optimization exercise. Furthermore, incomplete knowledge about the underlying economic processes, and uncertainty about the future path of exogenous variables (e.g. world prices), would most probably necessitate updating the optimization exercise after each new period. What this means in an *ex post facto* or historical context is that the choice of period to furnish the data for the optimization exercise is likely to have a noticeable effect on the calculated shadow prices for all the periods. The search for an absolute or 'correct' set of shadow prices, as some writers seem to have advocated in the literature, thus becomes problematic.

The calculation of base point prices as shadow prices or the duals in an intertemporal optimization exercise may also face practical problems due to lack of available information and the paucity of the historical database. This may explain why none of the empirical studies reviewed in this book has applied this method. As a practical alternative, the use of international border prices as shadow prices has been advocated in some of the literature. This is based on the argument that, from the point of view of any sector or institution, the opportunity cost of selling its output to (or purchasing its inputs from) other sectors of the economy is the foregone opportunity of making such exchanges with the rest of the world. Hence, to estimate the terms of trade effect arising from the 'distorted' prevailing market prices, the correct base point prices would be the international border prices. This solution, too, however, seems to suffer from similar problems to those discussed above. First, the international prices, particularly those of primary commodities, are generally much more volatile than domestic prices. This implies that base point prices have to be updated continuously, so that they reflect the terms of trade at current border prices. For example, measurement of the terms of trade effect in which base point prices are assumed to be the world prices of ten years ago is of little analytical or practical value, because the latter set of prices in no way reflects the opportunity cost of the resources exchanged at present.[7] With an annual updating of the base point prices, the calculated terms of trade effect in each year shows the consequence of the divergence of domestic prices from the hypothetical situation in which current world prices would prevail. Obviously, this measure could vary from year to year as a result of changes in world prices, even if everything else in the domestic economy remained the same.

The second set of problems associated with using world prices as base point prices is that this implicitly assumes that the country in question can exchange if necessary its entire intersectoral resource flows at current world prices. This may not be a realistic assumption, particularly in relation to the larger countries in our case studies, such as India and China. A related problem is that it does not seem plausible to regard world prices as base point prices in calculating intersectoral resource flows when we know that, if resources were actually valued at international prices, then the quantities exchanged, and, indeed, the structure of the economy as a whole, would be entirely different. An obvious example, looking at the problem from the real side, is when agricultural production and agricultural marketed surplus are sensitive to price changes. Many other examples could be found by looking at the decomposition of financial flows as depicted on the right-hand side of Equation 2.7 above. For example, government investment in developing countries forms an important source of inflow into the agricultural sector. If this investment is made possible by tariff revenues of the government, then it would be unrealistic to assume that the flow would remain the same even if the tariffs were removed. One may think of setting up a counterfactual exercise in

[7] Some of the results reported in Ellman (1975), where agricultural surplus in the Soviet Union is calculated at 1913 world prices, are an example of this type of exercise.

which not only prices, but also quantities were allowed to change. This, however, takes us right back to where we started, i.e. deriving shadow prices as duals in an intertemporal optimization programme.

This is not of course to deny that world prices convey valuable information which should be taken into account in deciding on an appropriate base point price system in the valuation of the terms of trade effect. It is, however, important to be aware of the problems involved in treating world prices in any particular year as the correct base point price system. The search for an absolute or correct set of base point prices has led some economists to resort to the labour theory of value as expounded by the classical political economists. According to this theory, under competitive conditions and assuming full mobility of factors of production, market prices would reflect labour time embodied in different commodities if labour were the only factor of production. On this basis, a number of Chinese and Soviet economists have tried to measure base point prices according to the labour embodied in the commodities exchanged between agriculture and the rest of the economy. This, however, raises a number of objections. First, with the introduction of capital as a factor of production, relative prices in competitive conditions would correspond to labour embodied in the exchanged commodities only under the very unlikely condition that the composition of capital (capital/labour ratio or the degree of mechanization) happens to be the same in the agricultural and non-agricultural sectors.[8] Secondly, even if we assume labour to be the only factor of production, there still remains the problem of comparing labour of different qualities, under different organizational conditions and work intensities. This latter problem is particularly important in developing countries with a low factor mobility between sectors, and with substantial surplus labour and disguised unemployment in the agricultural sector. To deal with these problems, Chinese economists have tried to devise equivalence schemes for labour spent in the agricultural and non-agricultural sectors. This is usually done by dividing the agricultural labour hours by a fixed coefficient which is meant to reflect the equivalence scale for labour hours between the two sectors. The estimation of this coefficient, however, is inevitably affected to a large extent by the subjective judgement of the researcher. The range of the values assumed in empirical work has varied between 2 and 8, which clearly reflects the problems associated with this approach to the terms of trade measurement.[9]

The conceptual and practical difficulties in establishing an 'absolute' or 'correct' set of base point prices mean that it may be necessary to adopt a more pragmatic approach to the measurement of the terms of trade effect. To heed the lessons from theory, it would be plausible to adopt as base years those years in which the economy shows least signs of disequilibrium, or, in other words, to opt for what

[8] This is the familiar transformation problem discussed in Marx (1977, pp. 142–208) and further elaborated in later literature (see e.g. Seton, 1957).

[9] For a critical review of the work of Chinese economists in measuring the terms of trade effect, see Sheng (1992).

Ishikawa (1967*a*) has referred to as 'normal' years. Years of major harvest failures, industrial disruptions, and severe foreign exchange shortages are clearly inappropriate as base years. Similarly, years of major policy change, when the economy has not had the time to adjust to the new policy regime, have to be avoided. The relative prices thus arrived at should be compared with world prices and the sensitivity of the results to the changing international relative prices should be assessed. Even with these precautions, however, it would be pointless to presume that we have arrived at a correct set of base point or reference prices and that our measure of income terms of trade is in some sense the correct or true one. Under these circumstances, it would be more plausible to regard our measure of real net intersectoral resource flows as an index of net surplus transfer, indicative of the changes in surplus flow over time rather than a correct or true measure of the absolute value of resource flows.

The main interest of all of the empirical studies considered in this book is in the behaviour of intersectoral resource flows over long periods of time under different institutional set-ups and policy regimes. Rather than being concerned with the absolute value of real intersectoral resource flows, we are essentially interested in the change in the resource flow index over time. Though the direction of net resource transfers is also of obvious interest in our analysis, due care must be taken before we can make reliable inferences as to the direction and absolute magnitude of net surplus transfer. This involves checking the sensitivity of the results to the use of different base year prices, in addition to ensuring that the base years come as close as possible to 'normal' years.

2.6 Summary and Conclusions

In this chapter we set up an accounting framework for intersectoral resource flows which will act as a unified framework for the comparison of country case studies in the following chapters of the book. The accounting framework was constructed on the basis of a simple social accounting matrix with the minimum entries necessary to represent the intersectoral resource flows between agriculture and the rest of the economy.

The accounting framework was used to formulate different notions of agricultural surplus and demonstrate various decompositions of net surplus flow between sectors, both from the real and the financial sides. We also discussed the measurement of agricultural surplus in real terms and introduced a general formula for the measurement of the terms of trade effect. We identified two problems associated with the measurement of *real* net surplus transfer: namely, the choice of an appropriate price index for deflating financial flows; and the choice of the base year or reference point prices. We noted that the latter could significantly affect the magnitude and even the direction of net resource flows, and therefore that great care must be taken in the measurement and interpretation of the absolute value of real net intersectoral resource flows in empirical studies.

As we have noted above, the agricultural surplus can take different social forms, and there are various financial mechanisms through which intersectoral resource flows can take place. The multiplicity of mechanisms and social forms of intersectoral resource flow, and the diverse implications of each form for the utilization of the surplus, indicate the importance of a disaggregated analysis of intersectoral resource flows. We shall now use these financial and real-side decomposition formulae in discussing some of the main determinants of intersectoral resource flows, before proceeding to the examination of the empirical case studies in the following chapters.

3

Determinants of Surplus Transfer and Policy Constraints

3.1 Introduction

In this chapter we shall be dealing with the determinants of intersectoral resource flows and the constraints which they pose for government policy directed towards bringing about a particular pattern of surplus transfer. In recent years the issue of intersectoral resource transfer has taken a central place in the growing criticism of the industrialization policies adopted in the developing countries in the post-war period. It has been argued that the 'industrialization bias' of governments, mainly characterized by the pursuit of import-substitution industrialization policies, has led to a squeeze of investible funds out of the agricultural sector, where they have the highest returns. This is claimed to have retarded the growth of the agricultural sector and halted industrial growth as well (see e.g. World Bank, 1986). The argument has sometimes taken a political turn, with the 'urban bias' of the government being blamed for the distortion in resource allocation processes (see e.g. Lipton, 1977). The rationale for the criticism in both cases has been the allegedly much higher rate of return on investment in the 'capital-hungry' agricultural sector of these countries. While the issue of intersectoral resource transfer is central to such arguments, little has been done to substantiate these claims empirically, except for the general point that the import-substitution industrialization policies, by turning the terms of trade against agriculture, must have led to such a resource transfer.

This is not, however, an entirely satisfactory argument. The discussion in the previous chapter has revealed various mechanisms of intersectoral resource transfer, only one of which is the terms of trade mechanism. For example, it may be argued that import-substitution industrialization policies, by increasing government revenues in the form of import duties, can contribute to growth of government investment in agriculture or the provision of subsidies for strategic inputs such as fertilizers. The empirical evidence seems to suggest that the response of agricultural output to such strategic inputs is important, and indeed, adequate provision of the strategic inputs is itself an essential precondition for agricultural price responsiveness (Taylor and Arida, 1988). Furthermore, import-substituting industrialization need not necessarily lead to worsening terms of trade against agriculture. Productivity growth in the industrial sector, combined with greater demand for food as a result of greater industrial investment, may produce the opposite effect. This brings up the important issue of the constraints imposed by the empirically given structure of the economy on the government's ability to enforce particular patterns of intersectoral resource flows.

A plausible way to approach this problem is to analyse the determinants of intersectoral resource flows, under given initial conditions and development patterns, and then to examine the possibilities of government intervention to influence these determinants. In other words, it is only after examining the determinants of intersectoral resource flows under empirically given economic structures that one would be in a position to enumerate the policy options facing the government in instituting a particular pattern of resource flow. Similar considerations are necessary in analysing the economic implications and the desirability of such policy options as exist. One approach to this problem would be to set up a complete multisectoral model based on the social accounting matrix discussed in the previous section, and to perform simulation exercises under different policy assumptions. This kind of approach, however, apart from being impractical due to the lack of the necessary data for the historical country studies considered here, is also of limited value in addressing the type of questions involved in the analysis of the long-term and dynamic processes of intersectoral resource transfers. Some of these limitations have been highlighted in the criticisms levelled against the existing SAM-based computable general equilibrium models in the literature (see e.g. Bell and Srinivasan, 1984). The comparative static type of analysis conducted by such models at their present stage of development is not suitable for the study of long-term and dynamic processes of change relevant to the intersectoral surplus flow process. Furthermore, the existing models totally neglect the financial side, which forms an important aspect of intersectoral resource flows. In addition, if not totally neglected in such models, factors such as technological and institutional change which are accepted as playing a central role in the resource flow process are taken as given. The production organization in agriculture, which conditions both the generation of the surplus and the various forms that it takes, is usually treated in a cursory manner in the general equilibrium models, in which agricultural response is normally modelled as a simple function of relative prices.

Another approach may be to analyse the determinants of intersectoral resource flows with the help of the more aggregative analytical models of the dual economy discussed in the literature. As noted in Chapter 1, however, various of these models focus on particular aspects of intersectoral resource flows and, for analytical manageability, abstract from other aspects which may none the less be of significance in the flow of net agricultural surplus. For example, the models by Lewis (1954) and Jorgenson (1961) abstract from the flow of investment goods from industry or the modern sector to agriculture or the traditional sector. This practice is also followed in most of the dual economy models which were developed in later literature (Dixit, 1973). Chichilnisky and Taylor (1980) describe a general equilibrium two-sector model in which the terms of trade between agriculture and industry adjust to reinstate equilibrium in the wake of exogenous shocks or policy changes both in the short run and in the long run. They highlight the constraint that agricultural marketed surplus may pose for industrial growth in a market economy in the long-run version of their model. The assumptions of perfect capital mobility between the sectors and fixed technology, combined with the fact that they abstract from official and financial current and capital flows, make their model of limited use in analysing the intersectoral resource flow processes in the long run. Other recent papers by Sah and Stiglitz (1984, 1987) specifically address the question of agricultural surplus transfer through the terms of trade mechanism – as highlighted in the Soviet industrialization debate – in the context of a neoclassical general equilibrium model. By assuming the sales and purchases of each sector to be equal, they concentrate on the terms of trade as the sole mechanism of surplus flow. Their model is subject to similar shortcomings to those of Chichilnisky and Taylor, therefore, with the additional problem that they totally abstract from capital in their agricultural production function. This is a drastic simplification, considering the importance of capital in both agricultural development and intersectoral resource flows and, not least, its significance in resource flows in Soviet agriculture during the 1920s (see Ellman, 1992).

Another interesting formal model of intersectoral resource flows is provided by Ishikawa (1967*a*). Ishikawa's study is particularly valuable from our point of view as it is based on the stylized facts derived from his extensive empirical studies of Asian agriculture, which incorporate the sample of countries considered in our own case studies. The model developed by Ishikawa differs from those discussed above in that it concentrates precisely on those aspects of the resource flow process from which other models normally abstract. The main focus is on real intersectoral resource flows, of both consumer and producer goods, and particularly the flow of investment goods into agriculture, which Ishikawa regards as crucial for output and productivity growth in the sector. Ishikawa's model in turn abstracts from price effects, and, as in the other models, the financial side of the resource flow process is not considered.

Despite their shortcomings, the various two-sector analytical models in the literature still provide valuable insights into various aspects of the intersectoral resource flow process which we shall take into account in the following discussion. At this stage, however, we shall refrain from trying to analyse the determinants of resource flows on the basis of formal economy-wide model-building. To include all the possible determinants of surplus flow in a model which incorporates both the real and financial sides of the resource flow process would be impractical. Even if it were practical, it would be unlikely to provide any real insight into the resource flow process to guide our empirical case studies in later chapters, and the results in any event would very much depend on the closure rules and adjustment processes assumed. The adoption of appropriate closure rules, as well as the possibility of abstracting from unnecessary details, presupposes an investigation of the empirical experience of resource flows in the case of the countries under study. This is not to underrate the significance of formal economy-wide modelling as such. The usefulness of such formal modelling, however, depends in the first place on a realistic characterization of the behavioural relations of the model as well as an appropriate method of abstraction which retains the essential features of the economy in question in the model. The kind of comparative historical country study conducted in this book is therefore to be regarded as a precondition for the construction of sensible formal models. In this chapter we shall try to identify some of the basic determinants of surplus flow in the agricultural sector on the basis of the disaggregated formulae set up in the previous chapter. This should be regarded as a general framework for the empirical studies in the subsequent chapters rather than as a deductive model of the determinants of surplus flow as in some of the purely theoretical models in the literature. Our main emphasis will be on the notion of net resource transfer or the 'net finance contribution' of the agricultural sector. Other notions of agricultural surplus discussed in the previous chapter could be analysed in a similar manner, and their determinants turn out to be more or less similar to those of the net finance contribution of agriculture.

The chapter is organized in the following way. The next section concentrates on the determinants of surplus transfer from the real side. The financial mechanisms of resource transfer are discussed in Section 3.3, where we also discuss some of the economy-wide aspects of intersectoral resource flows and their policy implications. The conclusions of the chapter are discussed in Section 3.4.

3.2 Determinants of Resource Flows on the Real Side

Consider the following decomposition of net intersectoral resource flow, based on Equation 2.4, where all the variables are measured in real terms:

$$r = x_a - m_a = f_a - (c_{af} + c_{nf}) - (i_{af} + i_{nf}). \qquad (3.1)$$

The net surplus outflow from agriculture in real terms is equal to the value added in the sector (f_a) minus the consumption of the farm households ($c_{af} + c_{nf}$) and investment in the sector ($i_{af} + i_{nf}$), procured from both within the sector and outside. For the time being we shall abstract from price changes and assume that the financial side of the resource flow process adjusts passively to accommodate the commodity-side changes without a second-round effect on the latter. It may be assumed that the necessary adjustments are made by the government in the financial markets and in the non-agricultural sectors to ensure equilibrium in both the financial and the goods markets. Though we shall consider the likely effects of relaxing this assumption as we proceed, at this stage it is helpful to concentrate on the impact of some basic real-side variables on the surplus flow process.

Let us begin with some static aspects of the resource flow process. A basic precondition for a net surplus outflow from agriculture, as shown by Equation 3.1, is that the value added in the sector should be greater than the consumption of direct agricultural producers and the investment needs of agriculture for maintaining the current levels of production. This means that, in agrarian economies with high population pressure on land and low levels of labour productivity in agriculture, where agricultural output is hardly sufficient to cater for the basic consumption of the farm households, the possibilities for surplus outflow from the sector will be extremely limited. By implication, in a near-subsistence economy in which population growth and lack of absorption of that growth by other sectors of the economy create increasing population pressure on land, agriculture is likely to impose a growing financial burden on the rest of the economy. This, of course, is a direct outcome of the classical assumption of diminishing returns to labour under conditions of land scarcity, and assumes that technological progress in agriculture is independent of population pressure. With the further assumption of agricultural output being dependent on the level of consumption of agricultural labourers, as in efficiency wage theories (see e.g. Bardhan, 1979), this condition becomes even more stringent. With the introduction of factors such as population growth, technological progress, and intersectoral labour mobility, we have already entered the realm of dynamic aspects of intersectoral resource flow process. Additional dynamic considerations are introduced by the fact that the intersectoral resource flows take place in the context of a growing economy which is undergoing structural change. It is these dynamic considerations to which we now turn. But what a static reading of Equation 3.1 has told us is that, for a surplus outflow from agriculture to be at all possible, the agricultural sector should have attained at least a minimum threshold of labour productivity.

The Consumption Side

In a growing economy, what happens to the real net surplus flow from agriculture obviously depends on the differential rates of change of agricultural value added on the one hand, and the real consumption and investment in the sector on the other. We shall be treating the consumption and investment aspects in turn. Taking the consumption aspect first: as can be seen from Equation 3.1, it is the difference between the increments in value added and in the total consumption (both self-procured and purchased from other sectors) that matters. The purchased component directly increases the inflow of resources, as it is normally an important component of resource inflow to the sector (m_a), and the self-procured component acts indirectly through reducing the marketed surplus (x_a). We shall begin, therefore, by looking at the behaviour of total consumption in the farm sector.

The single most important determinant of the agricultural or farm sector consumption is, of course, the disposable income of the farm households. The latter consists of value added in the agricultural sector (f_a) plus a number of items on the financial flow accounts such as the inflow of factor income from non-agricultural activities (minus the outflow of factor income, e.g. net rents), the net current transfers, and the income gains through the terms of trade effect.[1] Of course, value added in the agricultural sector, or that part of it which accrues to the farm households, forms an important component of the agricultural disposable income, and hence the movement of these two variables in the long run is expected to be highly correlated. Thus, to start with, we assume that the value added in agriculture forms the entire disposable income in the sector, and introduce the impact of other sources of income in steps. With this assumption, and bearing in mind that the average propensity to save is expected to increase with the increase in income, we can conclude that the higher the rate of increase in value added in the agricultural sector, the higher the increase in surplus outflow from the sector will be, for any given level of investment in the sector.[2] Assuming a stable age composition of the agricultural population and a given labour force participation rate, this could be translated into a condition related to labour productivity in agriculture. Dividing both sides of Equation 3.1 by the agricultural labour force, we can restate the above condition thus: the higher the growth of productivity of labour in agriculture, the higher the increase in surplus outflow from agriculture is likely to be over time, for any given rate of investment in the sector.

Growth of labour productivity in agriculture can result from a combination of factors, such as increased capital intensity, innovations in production techniques, and organizational and institutional innovations. In agrarian economies with population pressure on land and high levels of labour surplus in agriculture, labour productivity can also be increased by the absorption of the surplus labour in other sectors of the economy – a phenomenon referred to by Ishikawa (1967*a*) as the 'surplus labour effect'. Clearly, the higher the contribution of organizational and technological innovations to labour productivity growth, the higher the potential for surplus outflow will be – as this would economize on the necessary investment to achieve a given rate of labour productivity growth. (We shall return to this point shortly, when discussing the investment side of surplus flow process.) If such innovations are combined with a simultaneous outflow of surplus labour from agriculture, the impact on surplus outflow would be magnified, as the consumption-inducing effect of the labour productivity gains would be lower – even a decline in total consumption under such circumstances is not inconceivable.[3] It follows that, in countries with high population growth rates and limited employment generation in non-agricultural sectors, either due to the slow growth of these

[1] The terms of trade changes can also affect farm household consumption through their effect on the valuation of farm sector assets, but here we are mainly concerned with their income transfer effect.

[2] Of course, this also depends on the distribution of income in agriculture, to which we shall return shortly.

[3] This was one of the assumptions of Lewis (1954), who suggests that surplus labour, in moving to the organized sector, carried along its wage basket. It would, however, be more plausible to assume that the consumption of those on whom the migrant labour relied hitherto would increase somewhat as a result of their higher per capita income levels.

sectors or to the capital intensity of modern industrial technologies, the possibilities of surplus extraction from agriculture are likely to become increasingly limited over time. As we shall observe in the empirical country studies, this situation seems to have dominated the resource flow process in countries such as China and India.

In general, then, one may assume that, in economies with fast rates of growth of labour productivity in the agricultural sector, either due to rapid growth of the industrial labour force absorbing the agricultural surplus labour or fast technological change in agriculture, the net finance contribution of agriculture to economic growth is likely to be increasing over time. This characterizes the case of dynamic interaction between agriculture and industry in the process of development discussed in Chapter 1, which is usually neglected in the comparative static and often short-term analysis of the general equilibrium models of intersectoral resource transfer. This process could be termed the 'natural' process of surplus transfer, in the sense that it forms an ongoing process in time, whereby growing surplus transfer can go hand in hand with the increase in agricultural consumption, and structural change in the economy as a whole. Under these circumstances, there are ample possibilities for government intervention to increase the outflow of surplus even further, by checking the growth of agricultural consumption. This brings us to the discussion of the consumption effect of the income flows on the financial side.

So far, in considering the consumption side of the resource flow process on the basis of Equation 3.1, we have been assuming that the entire agricultural value added accrues to the farm households and that it forms the only source of their income. As argued above, however, the farm household disposable income is affected by various other mechanisms of income transfer on the financial side. These introduce added dimensions to the analysis of consumption effects in the surplus flow process. To start with, not all of the value added in agriculture accrues to the farm households. The extent to which it is siphoned off in the form of factor income outflows such as rent and interest payments depends on the prevailing agrarian relations and the development of rural financial markets. For example, in agrarian economies dominated by large-scale absentee landlordism, a considerable share of the agricultural value added may flow out in the form of rents and interest on loans from the informal credit markets at usurious interest rates. This outflow is reflected on the real side in the fact that, for any given level and rate of growth of labour productivity in agriculture, the level and the rate of growth of consumption of the farm households would be proportionately less. It would, however, be a mistake to infer from this that agrarian relations of this type would necessarily imply a higher rate of surplus outflow from agriculture and a larger net finance contribution of agriculture in the long run. For these same relations may at the same time hinder the growth of labour productivity in agriculture by discouraging investment and technological innovations in the sector. In the long run, the retardation effect on value added in Equation 3.1 is likely to outweigh the consumption-reducing effect.

The second source of factor income flow, which may lead to a rate of growth of consumption in agriculture over and above that warranted by the growth of labour productivity in the sector, takes the form of wage labour income by members of farm households partly engaged in non-agricultural activities. The same applies to income transfers by migrant workers and income gains through the terms of trade effect. In a predominantly agrarian economy with sluggish growth of labour productivity in agriculture and booming non-agricultural activities, these sources of income can lead to a much higher rate of growth of per capita consumption in agriculture than of labour productivity, and hence a growing financial burden by agriculture on the rest of the economy. This type of boom cannot last long, however, as the food supply constraint and the inflationary pressures in the economy would eventually end the boom in the non-agricultural sectors. It is sometimes argued that this sort of income transfer would itself contribute to the growth of labour productivity in agriculture by furnishing the necessary funds for investment in the sector and, in the case of the terms of trade effects, by improving work incentives and providing further inducements to invest in the sector. For this proposition to be true, however, the existence of a host of other conditions related to agrarian relations, basic infrastructural investment, the availability of appropriate and profitable technological innovations, and the existence of institutions for the diffusion and efficient use of the new innovations, has to be ensured. As we will see in the case of India, for example, successful diffusion of the green revolution technology only took place in a few regions where these latter conditions prevailed, while in the majority of the states, where these enabling conditions did not hold, the new income transfers during the period of the 'new agricultural strategy' only helped to inflate consumption of the farm sector (see Chapter 5).

Of course, the increase in income transfers to agriculture need not necessarily be absorbed by consumption or investment in the sector itself. At high levels of agricultural productivity and income, and depending on the degree of development of financial markets in the countryside, increases in farm household incomes may create an equivalent outflow through private capital transfers in the form of voluntary private savings. As we will observe, this mechanism was at work in the later stages of development in countries such as Taiwan and Japan, where relatively high levels of agricultural productivity were already achieved and the network of financial institutions in the countryside was widespread. At the early stages of development, when these conditions are not satisfied, however, a major part of such income transfers is likely to be spent in the agricultural sector itself, most probably on consumer goods. Under these circumstances, it may be essential for the state to siphon off at least part of the new income transfers in the form of land taxes and other kinds of direct taxation, in order to prevent agriculture becoming a growing financial burden on the rest of the economy.[4] In doing so, however, due care must be taken to ensure that high taxes do not diminish work and investment incentives in agricultural production.[5] The best way of doing this is for the state simultaneously to make adequate investment in basic infrastructure for agricultural production and to help introduce a stream of innovations so that profitability and productivity growth in the sector will be maintained. As we will note in later chapters, the combination of these factors played an important part in the earlier stages of development in Japan and Taiwan, ensuring a net positive financial contribution by agriculture to other sectors of the economy.

Our discussion so far has neglected differentiation between farm households and its implications for intersectoral resource flows. The agricultural sector in developing countries does not normally consist of a homogeneous peasantry, and this introduces a further consideration in the analysis of surplus flow on the consumption side, namely, the impact of the distribution of the agricultural incomes on surplus flow. Given the higher savings propensity amongst the high income farm households, it may be postulated that the more skewed the distribution of income in the agricultural sector, the lower the total consumption and the higher the net surplus outflow will be, at a given level of agricultural output (Ishikawa, 1967*a*). In a dynamic context this proposition can be restated in terms of growing differentiation amongst the peasantry and growing surplus outflow from the sector. This process of differentiation can be regarded as one aspect of what Marx (1976) has referred to as the process of primitive accumulation. However, for this to be a continuing process over time, that is, for differentiation to lead to an ongoing process of growing surplus outflow, it must be combined with technological advances which can give rise to increasing productivity of labour in agricultural production. Otherwise, the increase in the surplus outflow through income redistribution mechanisms would have a once and for all effect, as it is not the result of an increase in the productive potential of agriculture. By reducing the productivity of agricultural labourers in the poorer sections of peasant households, as emphasized by efficiency wage theories, it may even have a negative effect on surplus flow in the long run.

On the consumption side, therefore, the growth of labour productivity in agricultural production seems to be central in securing the possibility of a growing surplus outflow from agriculture in the long run. This in turn requires expanding fixed and working capital investment in the sector. The net result on intersectoral

[4] This is particularly relevant to income gains through terms of trade effect, as such income gains are likely to accrue predominantly to the rich farmers and landlords who control the major share of the marketed surplus. The curtailment of non-essential and luxury consumption by this group is likely to have much wider beneficial effects in financing accumulation in the economy as a whole (see e.g. Kalecki, 1970).

[5] It should be noted that land taxes could in themselves prove to be an important boost to work and investment incentives in agriculture, as they encourage an intensification in land use and discourage absentee landlords from keeping land idle.

resource flows depends on the investment requirement per increment of agricultural output. This necessitates an investigation of the investment aspect of intersectoral resource flows on the real side, to which we shall now turn.

The Investment Side

The second major determinant of agricultural surplus, on the basis of Equation 3.1, is the level of agricultural investment relative to output. As can be seen, given the level of consumption in the farm sector, the higher the incremental capital output ratio, the lower will be the net finance contribution of agriculture to economic growth.[6] Ishikawa (1967*a*) emphasizes this factor as a major determinant of surplus flow in selected Asian countries, where the need for heavy investment in irrigation implies a high incremental capital output ratio, and hence a net resource inflow into agriculture at the early stages of development. To the extent that capital intensity of production techniques could be varied in the agricultural sector, it would be possible for government intervention to influence the magnitude or direction of net resource flow through the choice of production techniques. This is particularly important in the case of the countries studied in this book, where the government plays an important role in basic infrastructural investment in agriculture, especially in irrigation. The adoption of more labour-intensive technologies for such investment, which utilize the abundant supply of surplus labour within the agricultural sector, is obviously the optimum choice from the point of view of intersectoral resource flows as well.[7] In a growing economy, however, the marginal product of capital in agriculture is bound to decline in the long run in the absence of technological progress and under conditions of land scarcity. This is the familiar law of diminishing returns to investment in agriculture, forcefully argued by Ricardo and other classical economists. As in the case of consumption–income flow balances discussed above, therefore, an important factor on the investment side in the long-term behaviour of agricultural surplus flow seems to be the pace of technological progress and increased efficiency of production inputs in agriculture.

[6] A point to be noted here is that, as shown in Equation 3.1, investment refers to both the purchased and self-produced components of investment ($i_{af} + i_{nf}$). However, as the self-produced component (i_{af}) forms part of the value added in agriculture, its changes would be neutralized by equal changes in agricultural value added (f_a). One may therefore subtract i_{af} from both investment and value added, and take investment to mean the purchased component of investment (i_{nf}) and value added to be net of self-procured investment ($f_a - i_{af}$). This only holds under the assumption of surplus labour – otherwise, the self-procured component of investment may lead to a reduction in the marketed surplus of the agricultural sector.

[7] This depends on the reaction of farm household consumption to the wage labour income inflow from such investments. For the cut in the gross inflow of resources brought about by the adoption of more labour-intensive technologies not to be neutralized by increased consumption in the farm sector, it is important to ensure that the new investment leads to high rates of increase in yields and labour productivity in agricultural production. This is an added reason for emphasizing the significance of technological innovations in agricultural surplus transfer.

In agrarian economies with high population density and scarcity of land, the growth of output and labour productivity in agriculture hinges on the introduction of land-augmenting technological advances. These consist of a combination of, on the one hand, fixed investment in land improvement and irrigation, and, on the other, the introduction of the new biological technology – new high yield varieties of seeds and biochemical nutrients and pesticides – the so-called 'seed-fertilizer technology' of the green revolution. These technological advances require increased investment in agriculture and substantially increase the purchases of new producer goods from outside the sector, as well as encouraging labour absorption in agricultural production. Despite this increased labour absorption, however, surplus labour in agriculture would inevitably lead to an increase in the productivity of labour. Similarly, though the new seed-fertilizer technology substantially increases the capital requirements of agricultural production, the empirical evidence suggests that at the same time it reduces the marginal capital output ratio in the sector (Mellor, 1973).[8] As we will see in the empirical case studies, the successful application of the new biological technology plays an important role in increasing output per unit of purchased new producer goods inputs in countries where agriculture makes a net positive finance contribution to the rest of the economy.

Of course, the efficient use of the new biological technology requires extra investment in complementary and supporting activities, as most of the literature in this field emphasizes (see e.g. Hayami and Ruttan, 1971; Barker and Winkelmann, 1974). This consists of increased investment in the research and development necessary for the adaptation and efficient use of the new technology, increased investment in land and water development, and the creation of new institutions for providing technical inputs and services to farmers.[9] This is likely to involve a substantial capital absorption by agriculture in the early stages of the introduction of the new biological technology, and may suggest a decreasing capital intensity of agricultural investment in successive stages.

Another important aspect of technological progress is the reorganization of agricultural production. Such reorganization can reduce the incremental capital output ratio and lead to a higher net finance contribution of agriculture to economic growth, by increasing the effectiveness of capital investment in agriculture. The role of institutional and organizational factors in the diffusion and

[8] According to Mellor (1973, p. 12), the empirical evidence from India suggests that 'in areas where they are well adapted, the increase in gross value of output of dwarf wheat and rice varieties is typically four times as great as the increase in cost of inputs purchased from the non-agricultural sectors.'

[9] The role of these complementary factors in the efficient use of the new technology is well demonstrated in the literature by the wide divergence in yields from the same technology under various conditions (see e.g. Barker and Winkelmann, 1974). A further characteristic of the new seed-fertilizer technology, which calls for direct state intervention for rapid diffusion of the technology, is the increased variability – and hence risk – associated with its adoption. Under these conditions, input subsidies and the provision of some type of insurance scheme at the early stages of the diffusion of the technology may be necessary.

efficient utilization of the new land-augmenting biological technology has also been increasingly emphasized in the literature (see e.g. Griffin, 1974; Ahmed and Ruttan, 1988; and Boyce, 1987). On the other hand, organizational innovations can help to increase on-farm production components of agricultural investment (i_{af}) and consumption (c_{af}), by more effective utilization of surplus labour and other internal resources of the agricultural sector. As can be seen from Equation 3.1, this latter factor would have the effect of concomitantly increasing agricultural output (f_a) together with the self-procured part of consumption and investment in the sector, and hence increasing the net resource outflow by reducing the purchases of the agricultural sector from outside for any given level of agricultural output. An example of this type of institutional innovation is the communal organization of agricultural production in China, which was instrumental in making effective use of surplus agricultural labour (see Chapter 9).

To sum up: in a long-term perspective, a basic determinant of agricultural surplus flow on the real side appears to be the rate of technological innovation and productivity improvement in agricultural production. In technologically stagnant agrarian economies, the agricultural sector would eventually become a growing financial burden on the rest of the economy. This may result from the natural growth of population and increasing population pressure on land, with diminishing returns, eating into the marketed surplus of agricultural products on the consumption side; or it may result from the growth of demand for food in non-agricultural sectors leading to increasing inflow of investment goods into agriculture, with diminishing returns. As noted above, there are various other intervening mechanisms which affect the magnitude and direction of resource flows, such as income distribution amongst the farm households, stratification of agricultural producers, and various channels of income transfer on the financial side. However, as long as these other mechanisms are not combined with technological progress and a continuous improvement of efficiency of resource use in agriculture, their effect in increasing the net product or net finance contribution of agriculture will not be long-lasting. It may be stated that, while technological progress and improved productivity in agriculture determine the potential for surplus outflow from the sector, other factors on the financial side of the resource flow account determine the form that this surplus takes and its actual magnitude. This statement, however, is only valid to the extent that one does not lose sight of common factors which affect both sides of the resource flow accounts, and the possible effects of financial flows on technological change itself. For example, as we noted above, the land tenure system in agriculture exerts a considerable impact on the financial mechanisms of resource flows, but at the same time it is expected to be of crucial significance to technological change in agriculture. Other mechanisms of financial flow such as the government's current and capital transfers and its credit policies, as well as the terms of trade effect, may also affect technological change in a significant way.

3.3 The Financial Side and Policy Constraints

On the basis of Equation 2.7, the financial side of the agricultural resource flow equation can be written as:

$$r = x_a - m_a = (f_a - y_f) - (k_{fg} - k_{gf}) - (k_{fo} - k_{of}) - (t_{fg} - t_{gf}) - (t_{fo} - t_{of}) - TT, \quad (3.2)$$

where the right-hand side variables, that is, the financial flow variables, are denoted in real terms, i.e. deflated by the agricultural sales price index, and TT is the income gains in the agricultural sector arising from terms of trade improvements. On the basis of this equation, the real net finance contribution of agriculture can be decomposed on the financial side into factor income flows, capital and current transfers on private and official accounts, and the terms of trade effect. To discuss the financial mechanisms of resource transfer, we have to remove the assumption of passive financial equilibrium made above. Instead, we begin by assuming a fixed magnitude of net resource flows determined on the real side. This is a simplistic assumption – for financial flows, as discussed in the previous section, can exert an important influence on the real side as well – but it is necessary at the outset for clarity of exposition.

As can be seen from Equation 3.2, a given level of agricultural surplus (r) can be extracted through various mechanisms. Government policy can strongly affect the relative magnitude of the flows through different channels, with important consequences for the process of development. For example, a land reform programme which reduces the component $(f_a - y_f)$ by removing absentee landlordism may be supplemented by an equal amount of land taxes to keep the surplus transfer intact. The result, however, would be far-reaching in terms of the utilization of agricultural surplus. Similarly, a change in the magnitude of surplus transfer through the terms of trade effect (TT) has its counterpart in an opposite sign in other financial items, for a given level of r. An attempt to increase the surplus flow through taxation of the farm sector or terms of trade changes, in the face of constraints set on the real side, would lead to a compensating reduction of flow through voluntary savings and financial channels $(k_{fo} - k_{of})$.

The interdependence of financial flows characterizes the limits that the given level of real resource flow (r) sets for policy – a rise in one item should be compensated for by a decline in another financial flow. Of course, the assumption of a given level of r, entirely determined on the real side of the economy, is not realistic. Different patterns of financial flows have different implications for agricultural development, and hence for the magnitude of resource flows. For example, an increase in the flow through terms of trade effect would change the supply of agricultural products, income distribution, and probably even the choice of technique in the sector. These would in turn affect the magnitude of total resource flow, as discussed above. The significance of these interactions is a matter for empirical research, but it should not be difficult to see that, in the absence of dynamic forces such as growth of productivity of labour and capital arising from technical progress, other types of influence would have a short-term and limited effect.

The above discussion has mainly concentrated on the agricultural side of intersectoral resource flow accounts, assuming that other sectors respond accordingly to maintain equilibrium in the financial and commodity markets. This can give rise to a one-sided view of the determinants of intersectoral resource flows. As noted in Chapter 1, the dynamic interactions between agriculture and industry can play an important role in the resource flow processes. For example, as seen above, the absorption of surplus agricultural labour by industry is an important determinant of the finance contribution of agriculture to economic growth. Similarly, industrial growth is necessary for the provision of the increasing amount of producer goods required by a technologically progressive agriculture. On the demand side, also, the growth of the non-agricultural sectors is essential for the absorption of the agricultural marketed surplus in an economy in which a technologically dynamic agricultural sector makes a growing net finance contribution to the rest of the economy. This condition becomes particularly stringent with the low income elasticity of demand for agricultural products, which gives an added significance to the absorption of surplus agricultural labour in non-agricultural sectors (Mundlak *et al.*, 1974). In the absence of an adequate rate of increase in demand, excess supply of agricultural products leads to a decline in agricultural prices which can inhibit investment and technological innovations in the sector. A more important aspect of the management of demand for agricultural products is to ensure stability of agricultural prices. Price stability is particularly essential, as it reduces the risks involved in the adoption of new technologies by the farmers, and it leads to a faster diffusion of those technologies. Another important reason for ensuring a certain degree of price stability for agricultural products is that it helps to prevent the propagation of shocks, which may originate in the agricultural sector on the supply side or in the industrial sector on the demand side, into deep and prolonged recessionary episodes (Kaldor, 1975, 1976). While government buffer stocks can play an important role in price stability in the short run, the co-ordinated growth of the industrial and agricultural sectors and the absorption of agricultural surplus labour by other sectors are essential for agricultural price stability from a long-term perspective.

3.4 Conclusions

In this chapter we have discussed some of the basic determinants of agricultural surplus flow on the basis of decomposition formulae developed in Chapter 2. Instead of setting up a formal theoretical model based on axiomatic assumptions about the behaviour of basic economic variables, we have adopted a more descriptive method whereby we could identify the basic technological and social conditions for the possibility of extracting a surplus from agriculture, and examine the different forms that the surplus can take and its possible basic determinants. This approach was adopted on the basis of the premisses that attempts to set up one general analytical model which could incorporate the various intersectoral resource flow processes in our different country case studies would be fruitless, and that our knowledge about the stylized facts of intersectoral resource flows is not yet adequate for fruitful formal treatment. The possible basic determinants of surplus transfer discussed in this chapter should thus be regarded as a guiding framework for the conduct of our empirical case studies and as working hypotheses to be verified by the empirical findings of the subsequent chapters.

From a long-term perspective, a basic determinant of agricultural surplus is likely to be the pace of technological progress and improved efficiency of production inputs in agriculture. In the absence of technological improvements, the natural growth of population would entail a declining productivity of labour in agriculture and a growing financial burden of agriculture on the rest of the economy. Attempts to reverse this process by increasing investment in agriculture, with given technology, would add to the financial burden of agriculture, due to the declining marginal efficiency of investment under conditions of land constraint. From this we may deduce two further important determinants of the agricultural surplus. The first is the level and the rate of growth of population. Other things being equal, a country with a greater population pressure on land and higher population growth rates will have a lower potential agricultural surplus and will find it increasingly more difficult to generate such a surplus. The second important determinant is the rate of absorption of agricultural surplus labour in other sectors of the economy. It can be said that, *ceteris paribus*, a country that exhibits a higher rate of absorption of agricultural labour by the non-agricultural sectors is likely to benefit from a higher rate of growth of net financial outflow from agriculture. This is partly due to the surplus labour effect, whereby the transfer of labour would automatically increase the productivity of labour in agriculture, and hence the potential surplus. This can have further beneficial effects by increasing the retained surpluses of the farm households, which facilitates the adoption of new technological innovations. In a country with a technologically progressive agriculture, the labour surplus transfer is also instrumental in generating an adequate rate of growth of demand for agricultural marketed surplus.

We have also considered the role of other factors such as distribution of income, and particularly choice of technique in agricultural production, as likely determinants of the net finance contribution of agriculture. In the absence of technological progress and continuous improvement in the efficiency of resource use in agricultural production, however, the effect of these factors is likely to be limited and of a short-term nature.

Considering the financial side of the resource flow process, we have seen how the net agricultural surplus can be decomposed into its different constituent elements. This decomposition demonstrates the various forms that agricultural surplus can take, and focuses attention on the effectiveness of economic institutions in facilitating surplus flow in different stages of economic development. It would be a mistake, however, to consider the financial-side constituents of agricultural surplus independently from its real-side determinants. As various examples have demonstrated, the interaction between the financial constituents of agricultural surplus and the real- or product-side determinants forms an important aspect of the intersectoral resource flow process. Government taxation of, and investment in, agriculture, factor income flows, terms of trade, etc., have important implications for real-side determinants such as income distribution, choice of technique, and consumption and investment in agriculture. These financial elements also interact with technological change in agriculture in an intricate way. Many of these interactions still remain relatively obscure and are subject to controversy in the development literature. It is hoped that a comparative study of intersectoral resource flows of the type conducted in this book will shed new light on aspects of these interactions.

The determinants of agricultural surplus discussed in this chapter define the possibilities of, and limits to, government intervention to influence the direction and magnitude of net intersectoral resource flows. They are partly rooted in the initial conditions inherited from the historical experience of growth, and partly result from the development strategies and the political constraints on government intervention in each country. Before conducting a country-by-country study of the determinants of surplus flow, therefore, it is essential to develop a comparative perspective on the initial conditions, the development strategies, and the overall patterns of structural change in our sample countries. This is the task of the next chapter.

PART TWO

Intersectoral Resource Flows in Historical Perspective

4

The Initial Conditions and Patterns of Development

4.1 Introduction

In this part of the book we shall examine the intersectoral resource flows of five countries during crucial stages of their development. The countries are China, Japan, India, Iran, and Taiwan. Though the number of case studies has had to be limited for the sake of manageability, nevertheless, the sample of economies considered is wide-ranging enough, in terms of initial conditions, resource availability, and development strategies, to allow useful generalizations to be drawn in a comparative study of this nature.

The intersectoral resource flows for different countries are examined for the following reference periods: India, 1951–70; Iran, 1963–77; China, 1952–83; Japan, 1888–1937; and Taiwan, 1911–60. In the case of India, the chosen period constituted a critical phase of development, when the government, in the immediate aftermath of national independence, attempted to achieve rapid industrialization within a planned mixed economy framework. The adoption of the new agricultural strategy in the last quarter of this period makes it of particular interest in relation to the analysis of intersectoral resource flows and industrialization strategy. In the case of Iran, the chosen period was one of rapid industrial growth, achieved by a high degree of protection of the domestic industry. The period is also of interest in that it includes the oil price boom of the early 1970s. According to neoclassical theory, such booms are expected to lead to a relative decline in the profitability of investment in the less protected traded goods sectors (in this case agriculture), and hence a shift of resources out of the agricultural sector – the so-called 'Dutch Disease' phenomenon. The study of intersectoral resource flows during this period would therefore be of interest not only with regard to the analysis of industrialization strategy and agricultural surplus transfer, but also in relation to the mechanisms of the Dutch Disease phenomenon. The 1952–83 period in China witnessed different experiments with central planning in order to achieve rapid industrialization in a state-dominated socialist framework. What makes this period of particular interest to the study of intersectoral resource flows is the common belief that the central planning institutions were mainly aimed at mobilization of resources within a predominantly agrarian economy for industrial accumulation. Apart from throwing light on the forms and functions of intersectoral resource flows in a rapidly industrializing centrally planned economy, the study of the Chinese experience in this period also helps to clarify the hypothesis with regard to the financing role of agriculture in the industrialization process. In the case of Japan, estimates for the post-war period are available, but here we shall concentrate on the pre-war experience, as that is the most relevant period in relation to the present-day developing countries. The years 1911–60 in Taiwan witnessed rapid agricultural transformation during two distinct phases of colonial and post-colonial development, which laid the foundations for the post 1960s phase of export-led industrialization.

Though the number of country case studies has been limited for practical reasons, they nevertheless form a varied sample in which each country can be taken as representative of a distinct class of economies with similar characteristics. Taiwan belongs to the group of newly industrializing economies (NICs) where, starting from a resource-poor agrarian economy, relative success in modernization of agriculture and diversification of industrial base has been achieved. Iran is an example of a resource-rich, oil-exporting developing economy which, during the 1960s and 1970s, benefited from rapid growth of foreign exchange revenues from the oil sector. China and India have various similarities in terms of initial conditions and economic structure, but the differences in their development strategies allow instructive comparisons with regard to the respective roles of policy and initial conditions in the resource transfer processes.[1] The experience of resource transfer in pre-war Japan, where resource outflow from agriculture is believed to have financed a major part of industrial accumulation, has often been quoted as an example to be followed by present-day developing economies. However, to be able to draw useful conclusions from the comparison of the experience of different economies, it is important to pay attention to the similarities and differences of the initial conditions and the development strategies which shaped the experience of resource transfer in each. The purpose of this chapter is to provide an overview of the initial conditions and the institutions which have played an important role in the intersectoral resource flow process in the country case studies.

The chapter is organized in the following way. In the next section we discuss the initial conditions, in terms of resource endowments and the productivity of resources in each country, with particular regard to the agricultural sector. Section 4.3 examines the institutional and organizational characteristics of the countries, as well as the development strategies and those aspects of government policy which are likely to have had an important bearing on the resource flow process. Section

[1] Ellman (1975) also provides estimates of intersectoral resource flows for the Soviet Union for the period 1928–32. As we have already included China as an example of a centrally planned economy, and since estimates for the Soviet Union are not comparable in detail to the other countries discussed here, we have not included this case study. The conclusions of Ellman's study, however, give further support to the conclusions reached in the present study (see also Ellman, 1992).

4.4 discusses the overall pattern of structural change in the countries under study during their respective reference periods, and examines the likely implications of different patterns of structural change for agricultural surplus transfer. The chapter is ended with concluding remarks in Section 4.5.

4.2 Resource Endowments and Factor Productivities

In terms of natural resource endowments, the initial conditions in all the economies in the sample, with the exception of Iran, were, by and large, similar. They could be broadly characterized as densely populated, with limited new agricultural land frontiers, and no significant endowments of exportable mineral resources. The case of Iran is distinguished by its large oil export surpluses. Net revenues from the oil sector financed more than 70 per cent of gross investment in Iran over the reference period. Like the other four countries, however, labour was the surplus factor in the case of Iran too. This was reflected in the relatively large share of agricultural labour force to be seen initially in all the economies. The share of agricultural labour force in the total was 83 per cent in China (1952), 72 per cent in India (1951), 54.0 per cent in Iran (1960), 70 per cent in Japan (1888), and 71 per cent in Taiwan (1911). In addition to the absorption of the natural increases in the labour force, the non-agricultural sectors in these economies also faced the task of absorbing a large share of agricultural labour in the process of development.

With regard to population growth rates, however, there were significant differences between these countries. The experience of Japan stands out from the other countries in our sample here. The annual rate of growth of population in Japan during the 1888–1920 period is estimated at about 1.1 per cent per annum, rising to about 1.5 per cent per annum during 1920–37 (Umemura, 1979). This compares with annual rates of growth which, on average, ranged between 2 to 3 per cent for the other countries in their respective reference periods. The average annual rate of growth of population in Taiwan increased from about 1 per cent during the 1911–26 period to 2.5 per cent during 1926–40 and 3 per cent during 1950–60 (Lee, 1971). Population growth rates in China, India, and Iran during their respective reference periods were 2 per cent, 2.3 per cent, and 3.1 per cent per annum respectively. This implies much higher rates of growth of labour force in countries other than Japan, which means that, to attain agricultural labour productivity growth rates similar to Japan's, the other countries had to maintain rates of growth in land yields that were at least 1 per cent per annum higher, or achieve higher rates of labour absorption in non-agricultural activities. Given the significance of the growth of agricultural labour productivity for intersectoral resource flows in the long run, as emphasized in the previous chapter, this can be expected to have introduced an important handicap for the other countries in terms of generation of agricultural surplus.

This points to yet another important difference in the initial conditions, which arises from the timing of industrialization in the different economies. The fact that the other economies, from the post-war period, had to use a much more advanced industrial technology than Japan in its reference period implies that they were likely to have a lower rate of labour absorption in their non-agricultural sectors for any given rate of capital accumulation and growth.[2] Of course, access to more advanced technology would generate a higher growth potential for output and labour productivity, particularly in the industrial sector, in the case of late-industrializing countries. In agrarian economies in which the inflationary barrier resulting from the food supply constraint limits growth, it can also allow a higher rate of industrial growth to be achieved – as the increased rates of labour productivity in modern industry would reduce labour absorption and thus the demand for food at any given level of industrial output. However, in the absence of the absorption of surplus labour from agriculture, the 'wage fund' constraint is likely to get increasingly tighter in an economy with high population growth rates and limited possibilities for the extension of cultivable land frontiers (see Chapter 10).

Of course, the rate of growth of yields and labour productivity in agriculture also depends on the initial levels of per capita national income and the resulting savings and investment capacity in the economy, which, in a predominantly agrarian economy, in turn depend on the levels of agricultural productivity at the beginning of the study period. With regard to per capita national income, and in particular productivity of labour in agriculture, there were important differences between the five economies under study. In the case of China and India, per capita national income in 1960 was below $400 measured in 1975 US dollar prices, while in Iran it was about $750 (Kravis *et al.*, 1982). This was largely due to large oil export revenues in Iran; excluding oil revenues, per capita national income falls to just over $400, which is a more appropriate indicator of labour productivity in the domestic economy. In the case of Japan and Taiwan, on the other hand, it is estimated that per capita national income, even at the beginning of the periods under consideration, namely, 1888 in Japan and 1911 in Taiwan, was more than twice that of the other three countries in the 1960s (Ishikawa, 1967*a*).[3] Given that the major part of the labour force in the countries under discussion was employed in agriculture, the above differences in per capita national income could also be taken as indicative of agricultural labour productivity differences in these economies. In fact, considering that labour productivity in the industrial sectors in China, India, and Iran in the 1960s may have been well above that of Japan and Taiwan in the last century, the differences in per capita national income could be indicative of even more glaring differences in agricultural labour productivity.

This proposition is supported by the existing evidence on yields and productivity of labour in agriculture in the initial reference periods in the different countries. According to the data provided by Kikuchi and Hayami (1985), land yields in

[2] This point is emphasized by Ishikawa (1967*a*), and is further elaborated in Section 4.4, where we examine the patterns of output and employment growth in the different countries in our sample.

[3] Excluding oil export revenues in the case of Iran.

Japan and Taiwan were more or less equal at the beginning of the present century, but since the land/labour ratio in Japan was higher, the productivity of labour in Japanese agriculture was about 20 per cent higher than in Taiwan.[4] The distance between these two countries and the other three countries in terms of the level of agricultural productivity at the beginning of their respective reference periods was, however substantial. Estimates of average yields in India indicate that, at the outset of its reference period, namely, in 1951, yields were a third of those in Japan in 1888 and only a quarter of those in Japan in 1900. However, since the land/labour ratio in India in 1951 was more favourable than in the earlier period in Japan, the gap between the two, in terms of agricultural labour productivity, was somewhat less glaring. According to our estimates, labour productivity in Indian agriculture in 1951 seems to have been about 58 per cent of that in Japan in 1888, and only 43 per cent of Japan's 1900 levels.[5] This also applies to China and, to some extent, to Iran as well. The estimated yields in Chinese agriculture in 1951 were noticeably higher than those in India in the same year, but since the land/labour ratio in China was well below that of India, the level of agricultural labour productivity in China was lower than that of India.[6] In the case of Iran estimates of agricultural productivity show that, while yields at the beginning of the study period were well below those of China and India, labour productivity was 40–60 per cent higher than in the other two countries, due to the relatively higher levels of land/labour ratio in Iran.[7] Compared to the cases of Japan and Taiwan, therefore, yields in Iranian agriculture seem to have been more than 50 per cent lower, but, in terms of labour productivity, the gap may not have been substantial.

As for the rates of growth of agricultural productivity over the study period, the sample countries show an equally varied experience, with Japan and Taiwan

[4] These figures refer to the years 1900 for Japan and 1913 for Taiwan. Output is measured in terms of wheat-equivalent units, and labour refers to male workers. For further details, see Kikuchi and Hayami (1985, p. 69).

[5] Estimates of yields and labour productivity (in wheat-equivalent units and per male agricultural worker) are provided in Kikuchi and Hayami (1985) and Hayami and Ruttan (1971) for Japan for 1888 and 1900, and for India for 1960. We have used the estimates of land, output, and labour growth rates in India provided by Chakravarty (1987) to extrapolate the productivity figures for India for 1951.

[6] See Patnaik (1988). This picture is also clearly demonstrated in the statistics of land and labour productivity for the two countries provided by the Food and Agricultural Organization (Agrostat, 1991) for the early 1960s period. According to this data, total cereal yields in China stood at about 1,277 kg/ha, averaged for the years 1961 and 1962, while the figure for India for the same years was about 938 kg/ha. Labour productivity, on the other hand, measured as cereal output per male worker in agriculture, stood at about 933 kg for India and about 683 kg for China. The gap between China and Japan, in terms of labour productivity during their respective initial periods, thus seems to have been much wider than that between Japan and India.

[7] For example, according to the FAO statistics (Agrostat, 1991), cereal yields in Iran, averaged for the years 1961 and 1962, were 868 kg/ha, as compared to 938 kg/ha and 1,277 kg/ha for India and China respectively in the same period. On the other hand, labour productivity, measured as cereal output per male worker in agriculture, was 1,318 kg in Iran, compared to 683 kg and 933 kg for China and India respectively.

performing better than the other countries in the sample. A detailed discussion of the achievement of each country in this regard will be provided in the country case studies. These differences in agricultural productivity growth to some extent depended on the initial levels of agricultural productivity and per capita incomes in general, which effectively set limits to the capacity of each economy to invest in, and renovate, the technological basis of agricultural production.[8] Other important explanatory factors refer to the organizational and institutional characteristics of the different countries and their development strategies and policies, which will be the subject of the following section. However, before ending this section, we should emphasize the significance of a further aspect of the initial conditions which could have played an important role in agricultural productivity growth rates, namely, the human capital factor, or the initial stock of skill and know-how in different countries. As we pointed out in the previous chapter, an important prerequisite for technological progress and efficiency of resource use in agriculture is the available stock of know-how, which is essential not only at the research and development stage, but also in the efficient utilization of the new technology.[9] The existing evidence suggests that, here again, the initial conditions in Japan in terms of human capital formation stood well above those in the other countries in the sample.[10]

4.3 Economic Institutions and Development Strategies

Amongst other important initial conditions relevant to the process of resource transfer were the institutional factors which affected the nature of policy inter-

[8] This signifies the importance of the existence of an external source of finance, such as the oil revenues in the case of Iran, in complementing national savings and, as we shall see, influencing the direction of net resource transfer.

[9] Taking formal education as an indicator of human capital formation, the available evidence for India, for example, suggests a strong correlation between agricultural productivity and educational levels in cross-section data, even after account is taken of variations in other economic features (see Sen, 1971). A similar correlation has also been found between the level of education of farmers and the speed of diffusion of new technology (see e.g. Desai and Sharma, 1966; Chaudhary and Maharaja, 1966; Shetty, 1968).

[10] Though formal education is only one dimension of the formation of human capital, it may nevertheless be taken as a reasonable quantitative indicator of the distance between the different countries in our sample with respect to human capital formation, or what Veblen (1932) refers to as the 'common stock of knowledge' of the society. In the case of Japan, a male literacy rate of 40–50% had already been achieved in the 1850s, and by 1914 virtually the entire population had attained functional literacy (Passin, 1965). In the case of India, on the other hand, the literacy rate in 1951 was only 17% of the adult population, and even by 1971 it was no more than 30%. The available information suggests that the literacy rate in China in the 1950s was even lower than India (see e.g. US Congress, Joint Economic Committee, 1967). In Iran the literacy rate for those over 10 years of age was only 30% in 1966 (CSO, 1966). The conditions in Taiwan in the early part of this century were even worse than in the other countries. However, in colonial times Taiwan benefited immensely from technical assistance from Japan, and by the time of Independence, in the post-war period, the educational levels surpassed those of the other countries in the sample, with the exception of Japan (see Lee, 1983).

vention by the government and organizational possibilities in each economy. As there exists a large body of specialized literature on these issues, we shall here only briefly point out the salient institutional features which played a significant role in the process of intersectoral resource transfer in the respective countries.[11] Where necessary, a more detailed discussion of these institutional factors is given in the individual country case studies.

In all of the economies in the sample, government played a major role in the process of mobilization and allocation of resources. In the case of pre-war Taiwan, a major determinant was the nature of Japanese colonial government, and its policy of using Taiwan as Japan's granary. Japan's colonial administration played an important role in infrastructural development and technological renovation of Taiwanese agriculture. The post-colonial Taiwanese government inherited a low-wage and highly productive agricultural sector. An important aspect of institutional change between the colonial and post-colonial periods was the land reform which removed absentee landlordism as a major agent of surplus extraction from agriculture. The ability of the government to maintain a low-wage economy, while extracting the agricultural surplus through taxation, after the land reform of the 1950s played an important role in mobilizing and directing national savings into industrial accumulation in the early decades of the immediate post-war period (see Chapter 6). Under Japanese colonial rule, Taiwan specialized as an agricultural commodity exporter, with the main part of its manufactured goods demand being imported, predominantly from Japan. Nevertheless, by the time of independence a number of important industries related to the agricultural sector, such as food and chemicals, had been established. For a short period during the 1950s, under the pressure of foreign exchange shortages, the government adopted an import substitution policy. This policy was soon abandoned, however, because of the smallness of the home market, and, from the early 1960s, promotion of export-oriented industries became the main focus of industrial policy, with a gradual shift from labour-intensive towards capital- and technology-intensive industries.[12]

In the case of China, the government faced the task of mobilizing savings in a poor agrarian economy with an extremely low degree of commercialization, through central planning. This involved the centralization of national surplus and its direct allocation by the state. A large part of savings was channelled into heavy industry. Important new organizational forms were also introduced, particularly in agriculture, which were instrumental in making productive use of the surplus

[11] For more detailed discussions and further references, see Lardy (1983) and Perkins and Yusuf (1984) for the case of China; Chaudhuri (1979) for India; Karshenas (1990*b*) and Katouzian (1981) for Iran; Ohkawa *et al.* (1979) for Japan; and Lee (1983), Lin (1973), and Ho (1978) for Taiwan. Ishikawa (1967) addresses production conditions in Asian agriculture, which covers all the countries in the sample, with the exception of Iran.

[12] Up to the early 1960s, light consumer goods industries were the leading sectors in manufacturing growth, but subsequently the capital goods and consumer durables became the leading sectors.

labour in the economy, as well as achieving a relatively fast rate of diffusion of new agricultural technology. The agrarian institutions in China during the study period underwent considerable transformations, characterized by collectivization and the replacement of the market mechanism by central planning directives during the Maoist period, and decentralization and the reintroduction of price incentives and a household responsibility system in the post-1978 period. This shift in the orientation of economic management seems to characterize two distinct phases of growth, indicating distinct objectives of economic planning. During the collectivization phase, the main objective seems to have been the acceleration of savings mobilization, capital accumulation, and productive utilization of the large pool of surplus labour in the Chinese economy. In the second phase, or the post-Mao decollectivization phase, the emphasis seems to have shifted to productivity improvements and attending to qualitative aspects of growth (see e.g. Patnaik, 1988). This shift in policy, as we shall see in Chapter 9, had important implications for the pattern of intersectoral resource flows as well. During the entire reference period, however, the overall industrial strategy remained inward-oriented, with a highly protected industrial sector mainly catering for the needs of the domestic economy.

The post-independence economy posed similar tasks for the Indian government, but different development strategies were adopted. As in China, the state played an important role in the allocation of investible funds, and initially great emphasis was put on heavy industry. Though the industrial priorities were somewhat modified from the mid-1960s, direct control of a major part of heavy industry and the system of industrial licensing meant that the government continued to play an important role in the sectoral allocation of investment in the economy. In contrast to the case of China, government intervention in the agricultural sector has been indirect and market-mediated, though its role in technological research and infrastructural investment and provision of credit has been important. As in China, the industrialization strategy in India was based on import substitution, whereby the manufacturing sector grew behind protective tariff walls catering mainly for the domestic market. A significant shift of policy during the study period which had important implications for the pattern of intersectoral resource flows was the adoption of the 'new agricultural strategy' from the mid-1960s. This involved a substantial shift of resources towards agriculture during the so-called green revolution period, as part of the attempt to renovate the technological basis of agricultural production.

In the case of Iran, the government played a dominant role in the economy through its control over oil revenues. In contrast to India, China, and Taiwan, *mobilization* of savings has not been a major preoccupation in Iran; the main task was rather the allocation of the already centralized funds in the form of oil revenues. This was attempted through direct investment by the government as well as provision of finance to the private sector through investment banks, within a mixed economy framework. As in China and India, the industrialization strategy of the government was based on import substitution, whereby the industrial sector grew rapidly during the study period under conditions of high protection and large subsidies by the government. As we shall explain in later country study chapters, however, the mere fact that the industrial sector was protected in these economies does not necessarily imply a bias in any particular direction for the terms of trade of the agricultural sector, as food imports were also under strict government control in all three countries. An important institutional factor which had a significant bearing on agricultural surplus flow in Iran during this time was the land reform programme which did away with large-scale absentee land-ownership. The strong reliance of the government on the oil sector for financing its expenditures also created an intrinsic weakness in the taxation system, with important implications for domestic resource mobilization and agricultural surplus flow. The availability of abundant supplies of foreign exchange from the oil sector also encouraged the adoption of a highly capital-intensive development path, both in industry and agriculture, with significant implications for the pattern of intersectoral resource flow (see Chapter 7).

In the case of Japan, also, the state played an important role, through the provision of finance, infrastructural investment, and science and technology research. As the agricultural sector had already achieved a high degree of commercialization during the period under consideration, the main channels of intersectoral resource transfer were market-mediated, though government taxation played an important role during the early Meiji period (see Chapter 8). An important aspect of agrarian organization in Japan at this time was the existence of an influential rural-based cultivating landlord class, which was receptive to new innovations introduced by government initiative. In contrast to the late-industrializing countries such as China, India, and Iran, where import-substitution industrialization involved a simultaneous development of heavy and light industries, industrial growth in Japan during the study period was based on light industries such as food and textiles. Exports of cotton textiles and raw silk formed a main source of foreign exchange revenues and constituted a major impetus for industrial growth in general. Another important aspect of government policy in Japan was the introduction of an effective system of taxation of agriculture. Land taxes, which formed a major source of government revenue in the early Meiji period, played an important part in the intersectoral resource flow process in this early period of development, when the rural financial markets were not yet adequately developed. A similar policy in Taiwan under the Japanese colonial administration distinguishes these two countries from the other countries in our sample, where direct taxation of agriculture remained rudimentary and, in some cases, virtually non-existent throughout their respective reference periods.

4.4 Patterns of Growth and Structural Change

As pointed out in the previous chapter, an important factor in the intersectoral resource flow process is the overall pattern of growth and structural change in the economy, particularly the relative sizes and growth rates of output and employment in the agricultural and non-agricultural sectors. Of course, growth and structural change in the economy is not exogenous to the intersectoral resource flow processes. The initial conditions in terms of resource availability, technological capabilities, and historically inherited institutions, combined with government policy and the international economic environment, give rise simultaneously to different patterns of both intersectoral resource flows and growth and structural change. In fact, in much of the development literature, and particularly in the debate on intersectoral resource flows reviewed in Chapter 1, it is the net intersectoral resource flows which are usually referred to as the exogenous policy variable in implementing different patterns of growth and structural change. We shall return to this issue in Chapter 10, where we examine the plausibility of this proposition. The purpose of this section is not to provide a causal analysis of the relation between growth and intersectoral resource flows, but rather to provide a comparative overview of structural change in different countries in terms of employment and output as an aid for further analysis in later chapters.

The structure of output and employment in the countries under study over their respective reference periods is shown in Table 4.1. As can be seen, there is a variety of experiences with respect to the speed of structural change and the relation between changes in the structure of output and employment in the different countries. China, India, Japan, and Taiwan have broadly similar initial patterns of output and employment, but exhibit significant diversities in the evolution of these patterns during their respective reference periods. The sectoral shares of output and employment in Iran in 1963 represented a more advanced stage of development compared to other countries in their respective initial periods, and structural change in Iran over the study period was much faster than the other countries. Before making a more detailed comparison of the individual country data in Table 4.1, however, it is necessary to mention one or two points with regard to the comparability of the data presented in the table. The first point to be kept in mind is that the definitions of sectors and categories in different country statistics are sometimes at variance.[13] The second point to be noted is that the period covered by the data for each country varies in length considerably, and hence the change in shares between the end-points does not translate directly into the speed of structural change.

The structure of output and employment in China in its first year indicate that,

[13] These differences are not so large as significantly to affect the overall trends reflected in the table, but the following important points need to be mentioned. 'Employment' in pre-war Japan refers to workers gainfully employed in different sectors, which tends slightly to underestimate unemployment compared to the definition in other countries (see Umemura, 1979). In the case of China, the output data exclude services other than commerce and transport, and the shares are expressed in terms of material product rather than GNP. It should also be noted that, in the case of Iran, both the employment and output figures refer to non-oil sectors, and the output shares are calculated as ratios of non-oil GDP. For other differences, see the notes to Table 4.1.

Table 4.1 Structure of employment and output in China, India, Iran, Japan, and Taiwan

	EMPLOYMENT (%)[a]			OUTPUT (%)[b]		
	Agriculture	Industry	Services	Agriculture	Industry	Services
China						
1952	83	7	9	58	23	19
1965	82	8	10	46	40	14
1979	70	18	12	37	52	11
1983	67	19	14	40	51	9
India						
1951	72	9	19	54	20	26
1961	70	10	20	52	19	29
1971	69	11	20	46	22	32
Iran						
1963	54	22	24	32	23	45
1975	36	33	31	14	29	57
Japan						
1888	70	12	18	54	16	30
1910	65	15	20	42	21	37
1937	45	23	32	19	31	48
Taiwan						
1911–15	71	29		48	27	25
1936–40	63	37		35	34	31
1950	56	44		36	19	45
1960	50	50		34	23	43

[a] Agriculture includes mining, and industry includes construction and utilities, with the exception of Iran, where mining is included in industry, and India, where employment in construction and utilities is included in services.

[b] Refers to value added shares in GNP (NNP in the case of Japan), with the exception of Iran, where figures refer to value added shares in non-oil GDP, and China, where they refer to shares in material products. Value added in services in China refers only to transport and commerce.

Sources: China: *Statistical Yearbook of China* (1987, 1991); India: Chaudhuri (1978); Iran: CSO (1968–80); Japan: Ohkawa and Rosovsky (1960), and Ohkawa and Shinohara (1979); Taiwan: Lee (1983, i, ii); Pang (1992).

compared to the other countries, it probably had attained the lowest level of development, with agriculture having the highest share in both output and employment amongst the countries in the sample. A striking feature in the pattern of structural change in China compared to the other countries is the marked difference in the change of sectoral shares between output and employment. While the share of industrial output increased rapidly over the study period, from 23 per cent in 1952 to more than 50 per cent by the end of the 1970s, the share of employment in different sectors remains relatively more stable.[14] This is, of course,

[14] It should be remembered, however, that since the output of services sector in China is restricted to commerce and transport only, the output share *levels* are not directly comparable to those of the other countries in the table.

indicative of much faster rates of growth of labour productivity in industry than in other sectors, and particularly in agriculture.[15] It would be a mistake, however, to deduce that the reason for the observed pattern of structural change is the faster rate of growth of labour productivity in industry than in agriculture. As we have noted in the previous chapter, productivity growth in agriculture in a surplus labour economy like that of China depends to a large extent on the rate of absorption of excess agricultural labour in other sectors of the economy. The Chinese government's policy of restricting both the spatial and sectoral mobility of labour played an important part in generating the observed patterns of structural change over the study period. This policy seems to have played a dual role in the process of structural change; on the one hand, by inhibiting the outflow of excess labour from agriculture it has retarded labour productivity growth in agriculture, and, on the other, by preventing the concentration of rural emigrants into the low-productivity informal industrial sector it has kept industrial labour productivity growth rates high. This process of absorption of excess labour in the informal non-agricultural sector tends to even out the rates of labour productivity growth in different sectors of the economy in countries which do not impose similar restrictions on labour mobility. With the relaxation of such restrictions in China from the late 1970s, we can observe a noticeable decline in the share of agricultural labour and a narrowing of the gap between the rates of growth of labour productivity in the agricultural and industrial sectors. However, despite the fact that the share of agricultural labour force in China declined from 83 per cent in 1952 to 67 per cent in 1983, the period nevertheless witnessed a considerable increase in the absolute number of agricultural labour force and in population pressure on land. Agricultural labour force rose from 173.2m. in 1952 to 277.9m. in 1971 and 311.4m. in 1983, an increase of more than 130m. at a time when land under cultivation was actually declining.[16]

The Indian economy during the study period was characterized by extremely slow rates of structural change, particularly in relation to the sectoral employment of the labour force. Starting from an initial economic structure which was remarkably similar to that of Japan and Taiwan, the Indian economy failed to undergo the same kind of structural changes as the latter two economies during the study period. The most significant structural change in India over this period (see Table 4.1) was the decline in the share of value added of agriculture by 8 per cent, which was largely captured by the increase in the share of services. What is particularly significant from the point of view of intersectoral resource transfer is the slow rate of decline of the share of agricultural employment, which fell by only 3 per cent

[15] This difference becomes even more noticeable if we consider the fact that the sector shares in the table are measured at current prices. Since the terms of trade over this period moved substantially against the industrial sector (see Chapter 9), the increase in the share of the industrial sector would be much more pronounced when measured in real terms.

[16] Agricultural labour here, as in Table 4.1, refers to the total labour force engaged in the primary sector.

during the two decades of the 1950s and the 1960s. As in China, given the slow rate of structural change and the relatively fast rate of growth of population, the absolute number of agricultural labour force increased rapidly during the study period, adding to the population pressure on land throughout the period. The number of agricultural workers increased from 101.9m. in 1950 to 137.8m. in 1960 and 167.3m. in 1970, leading to a decline in the availability of cultivated land per worker from 1.17 ha in 1950 to 0.84 ha in 1970.[17] As described in the previous chapter, such substantial increases in agricultural labour force, with limited cultivable land, are expected to have important implications for net agricultural surplus transfer.

The sectoral allocation of output and employment in the Iranian economy during its initial period, at least as far as the agricultural/non-agricultural shares are concerned, was closer to the structure of the Japanese economy in the 1930s and that of Taiwan in the 1950s, i.e. to the structure of these latter economies towards the end of their reference periods rather than the beginning. This apparently 'advanced' economic structure, however, was more a reflection of the peculiar pattern of development in Iran under the influence of its oil revenues than of the stage of development of the domestic non-oil economy. As we have already noted in the previous two sections, technological development of the Iranian economy, as reflected, for example, in its level of human capital formation and agricultural yields, was still well below the levels attained in the Japanese economy in the last two decades of the previous century. This point is further illuminated by the pattern of structural change over the study period. As shown in Table 4.1, compared to the rest of the countries in the sample, Iran achieved the fastest rate of structural change during its reference period. The share of agricultural labour force declined from 54 to 36 per cent, and the agricultural value added share in non-oil GDP fell by almost 20 per cent, from 32 per cent in 1963 to 14 per cent in 1975. This was brought about by the rapid rate of growth of the non-agricultural economy, propelled by the expenditure of fast-increasing oil revenues in the urban areas which attracted a sizeable number of migrant workers from the agricultural sector, largely absorbed in booming services and informal industrial activities. The large and expanding share of employment and output in services in Iran compared with the other countries in the sample was a prominent characteristic of this pattern of growth. Though modern manufacturing also showed rapid rates of growth in this period, its rate of absorption of labour was rather limited, and the majority of the labour force outflow from agriculture was attracted by construction, services, and low-productivity informal manufacturing (see Chapter 10). The rapid fall in the share of agricultural labour force meant that, despite the high rates of growth of population during the study period, agricultural labour force was following a declining trend during this period. It declined by more than 7 per cent during the study period, from over 3,749,000 in 1963 to about 3,469,000 in 1975.

[17] These figures are based on Chakravarty (1987, table 15, p. 113).

Given that this period was also one of increase in land under cultivation, these trends played an important role in sustaining the rate of labour productivity growth in agriculture, which, as discussed in the previous chapter, would, *ceteris paribus*, indicate a growing potential for a net surplus outflow from agriculture.

In the case of Japan, also, we can observe a rapid rate of structural change accompanying relatively high rates of overall economic growth. Central to the process of structural change was the strong industrial growth which induced the change in other sectors as well, notably the growth of services attending to the needs of a developing industrial economy, and the shift of labour out of agriculture to cater for the employment needs of these expanding activities.[18] Over the reference period the share of agricultural labour force declined from 70 per cent to 45 per cent, and the agricultural output share declined from 54 per cent to no more than 19 per cent. This pattern of structural change, combined with relatively slow rates of population growth, led to a considerable decrease in agricultural labour force in Japan over the study period as a whole. While between 1888 and 1937, the total labour force in Japan increased from about 22.8m. to 31.6m., agricultural labour force declined sharply, from about 15.8m. to 13.7m. This decline in the agricultural labour force, particularly in the interwar period, played an important part in sustaining a relatively high rate of labour productivity growth in agriculture in the face of cultivable land constraint (see Chapter 8).[19] This is in sharp contrast to the experience of countries like India and China, where increasing population pressure on land was an important condition to be taken into account in the discussion of the intersectoral resource flows.

Considering the 1911–69 period as a whole, the trends of structural change in Taiwan appear less impressive than those in Iran and Japan during their respective reference periods, but, compared to India and China, particularly in relation to the changing structure of employment, Taiwan exhibits more rapid structural change. It would be better, however, to consider the pattern of structural change in Taiwan during the colonial and post-colonial periods separately, as the disruptions of the Second World War, and particularly the huge inflow of migrant labour from mainland China in the post-war period, brought about major breaks in the process of structural change. The immigrant population added one million to a total population of merely six million in the post-war period. Considering the pattern of structural change in the colonial period, we can see that the share of agricultural labour declined during the three decades of the 1911–40 period from 71 per cent to 63 per cent, which was a steeper decline than that of either India or China in their

[18] For a detailed discussion of growth and structural change in Japan over this period, see Ohkawa (1979).

[19] Of course, the decline in agricultural labour force was not distributed evenly throughout the period. During the 1888–1906 period, agricultural labour actually increased slightly, from 15.8m. to 16.3m., and thereafter it continuously declined – with particularly high rates of decline during the 1915–30 period (see Umemura, 1979). During the early period of agricultural labour increase, however, land under cultivation was still growing, largely due to expanding land frontiers in Hokkaido.

respective reference periods. Despite the decline in the share of agricultural labour force, the absolute number of agricultural labourers increased throughout this period, from 1,106,000 in 1911 to 1,400,000 in 1940. This increase, however, was not large enough to lead to a substantial decline in land/labour ratio. Land/labour ratio actually increased from 0.60 ha per worker in 1911–15 to 0.68 in 1926–30, but thereafter it declined to 0.62 in 1936–40. In the post-war period, as a result of the massive inflow of migrant labour, the agricultural labour force jumped to 1,731,000 in 1950, and land/labour ratio fell thereby to 0.50 ha per worker. During the 1950–60 period, agricultural labour force continued its upward trend, but at a much slower rate compared to the colonial period. Agricultural labour force increased from 1,731,000 in 1950 to 1,754,000 in 1960, and the land/labour ratio declined slightly, from 0.50 to 0.48 ha per worker over the same period.[20] The extremely unfavourable and declining trends in land/labour ratio are expected to have put Taiwan at a disadvantage, in terms of the potential for agricultural surplus outflow, compared to other countries in the sample. As we shall see in Chapter 6, it was only due to fast rates of technological progress in agriculture and relatively low incomes and consumption of direct agricultural producers that a surplus outflow from Taiwanese agriculture was possible during the period under study.

4.5 Concluding Remarks

In this chapter we examined the initial conditions, in terms of natural resource endowments and the productivity of resource use determined by technological and institutional conditions and the initial capital stock, in the five economies which form our country case studies. We also considered aspects of structural change in these economies which could have a direct bearing on intersectoral resource flows. The main emphasis was on those aspects of the initial conditions and patterns of structural change which, on the basis of the analysis in the previous chapter, were expected to exert a significant influence on net agricultural surplus flow. To the extent that these conditions are inherited from the experience of development in the past, they can be regarded as constraints which limit policy choice in enforcing a particular pattern of intersectoral resource flows. However, these same initial conditions also evolve in the process of development, and hence an analysis of the actual determinants of surplus transfer requires a historical study of intersectoral resource flow processes in each country. In the next five chapters we shall examine the patterns and processes of intersectoral resource flows in each of the five countries separately, within the context of the overall comparative background provided in the present chapter.

[20] The absolute value of agricultural labour started to decline only from 1966, and it was not until 1973 that the share of labour force in industry first exceeded that of agriculture (see Oshima, 1986).

5

Intersectoral Resource Flows in India, 1951–1970

5.1 Introduction

In this chapter we examine the patterns and processes of intersectoral resource flows in India during the 1951–70 period. We have already discussed, and contrasted with the other countries in the sample, the initial conditions and the broad features of economic development in India over this period in Chapter 4. There we concentrated on the similarities and differences of the Indian economy in terms of its resource base, institutional set-up, and development strategies. That discussion forms the background against which we shall now analyse the patterns and processes of intersectoral resource flows in India, and is particularly helpful in drawing parallels and contrasts with the experience of other countries in subsequent chapters.

Further details of those aspects of Indian economic development which have a direct bearing on the resource flow processes will be discussed below where appropriate. However, to focus ideas and to introduce the theme which runs through this chapter and which also furnishes the link to the experience of other countries in following chapters, it would be useful at this stage to point out a factor which played a key role in the changing pace and pattern of intersectoral resource flows in India during the period under study; namely, the 'new agricultural strategy' adopted by the Indian government from the mid-1960s. This ushered in the era of the so-called 'green revolution', which had important direct and indirect implications for intersectoral resource flows.[1] The seed-fertilizer technology of the green revolution involved the introduction of a combination of new inputs into the agricultural sector, with obvious direct implications for the flow of resources into the sector. The new agricultural strategy also involved a considerable improvement in the agricultural terms of trade over the 1964–70 period, perceived by the policy-makers as an incentive device and also as a means of income transfer presumably meant to make the adoption of the new technology 'affordable' to the farming sector. This in turn had important implications for the intersectoral flow of resources through various channels, which will be discussed below. What

[1] The literature on the green revolution is voluminous. For useful reviews and further references, see Byres (1972), Griffin (1974), Rao (1975), and Chaudhuri (1978, ch. 5).

happened to the balance of resource flows, or the net surplus flow from agriculture, depended crucially on the productivity of new resources which were thus diverted to the farm sector.

The chapter is organized as follows. In the next section we examine the flow of resources between agriculture and the rest of the economy from the real side, i.e. from the point of view of the net commodity exchanges between agriculture and the rest of the economy. This is followed by an analysis of the determinants of resource flows in Section 5.3. In Section 5.4 we discuss the financial mechanisms of surplus transfer, and a brief summary and conclusions are provided in Section 5.5.

5.2 Resource Flows on the Real Side

There are various partial studies of agricultural resource transfer for India.[2] The only consistent time-series study of intersectoral commodity flows is that of Mundle (1981) for the period 1951–70. Mundle's estimates of real exchanges between agriculture and non-agriculture for consumer and producer goods are given in Table 5.1. Before considering these results, however, it is necessary to discuss certain aspects of Mundle's method of estimation in order to point out possible sources of bias in his estimates.

Mundle's estimates are based on a sectoral distinction between agriculture and non-agriculture. In estimating the agricultural sector's consumption, he meticulously separates agricultural and non-agricultural populations, and estimates per capita consumption in each group. This, however, does not solve the problem discussed in Chapter 2, namely, that agricultural households may be involved in household activities other than agriculture. In the face of difficult data problems posed by the correction for non-agricultural activities of agricultural households, and given that the scale of the error may not be large, this omission may be justifiable. But there seem to be a number of other sources of bias in Mundle's estimates. First, he has included grain processing as part of the agricultural households' activities. The reason for this, as Mundle points out, is the nature of the available official data, on the basis of which it is not possible to distinguish grain processing from grain production activities. He further argues that this may be justified by the fact that the processing of food grains is partly done within the agricultural households. This is not a plausible argument, and even if one has valid reasons for including grain processing within agriculture, then the number of agricultural households would have to be re-estimated to allow for the households in the grain processing industry. Given that this source of error is inevitable because of the nature of the available data, it would be important to note the direction of bias which it introduces into the intersectoral flow estimates. As the

[2] See e.g. Ishikawa (1967*a*) for estimates for the year 1951–2, Thamarajakshi (1969) for terms of trade estimates for 1950–1 and 1960–1, Krishna and Raychaudhuri (1980) and Mody (1981) for estimates of rural and agricultural savings respectively. Mody, Mundle, and Raj (1985) give a useful summary of the existing studies.

Table 5.1 Commodity purchases and sales of agriculture in India, 1951/2–1971/2 (bn. rupees, 1960/1 prices)

Year	Agriculture's imports			Agriculture's total exports				Agriculture's net exports	
				Mundle's estimates			New estimates		
	Consumer goods	Producer goods	Total	Consumer goods	Producer goods	Total	Total	Mundle's estimates	New estimates
1951–2	23.3	4.1	27.4	15.2	7.3	22.5	16.9	−4.88	−10.5
1952–3	29.1	3.6	32.7	16.7	7.6	24.3	18.2	−8.40	−14.5
1953–4	27.9	3.0	30.9	19.0	8.2	27.2	20.1	−3.76	−10.5
1954–5	25.1	3.3	28.4	19.0	9.2	28.2	21.2	−0.17	−7.3
1955–6	27.1	3.4	30.6	20.6	10.4	31.0	23.3	0.41	−7.2
1956–7	26.0	3.5	29.5	21.1	11.5	32.7	24.5	3.15	−5.0
1957–8	27.0	2.9	29.9	20.6	12.3	33.0	24.7	3.05	−5.2
1958–9	28.6	2.9	31.5	21.3	13.6	34.9	26.2	3.34	−5.3
1959–60	28.7	2.7	31.4	21.6	14.9	36.5	27.4	5.18	−4.0
1960–1	29.3	2.6	31.9	23.0	16.8	39.8	29.9	7.95	−2.0
1961–2	30.2	3.5	33.7	23.8	17.2	41.0	30.8	7.27	−3.0
1962–3	29.7	3.5	33.2	24.0	17.7	41.6	31.3	8.46	−1.9
1963–4	29.6	4.1	33.8	25.2	18.1	43.3	32.5	9.58	−1.2
1964–5	33.8	5.0	38.8	24.9	18.5	43.4	32.6	4.51	−6.3
1965–6	34.1	5.0	39.1	24.5	18.3	42.3	32.1	3.63	−7.0
1966–7	34.9	6.0	40.9	24.8	18.0	42.8	32.1	2.32	−8.8
1967–8	41.8	7.0	48.8	27.0	17.9	44.9	33.7	−3.89	−15.1
1968–9	41.9	7.9	49.8	27.3	17.9	45.2	33.9	−4.59	−15.9
1969–70	43.8	9.3	53.1	28.3	17.8	46.1	34.6	−6.96	−18.5
1970–1	43.9	10.7	54.6	29.4	17.3	46.7	35.3	−7.84	−19.3

Source: Mundle (1981).

output of the grain processing industry is included in the total sales of agriculture, while the consumption of households with grain processing as their major activity is not included as a purchase, Mundle's estimates of net surplus flow are likely to have an upward bias towards resource outflow from agriculture. Furthermore, given that, with the increasing commercialization of agriculture and the growing division of labour in the economy, grain processing activities are likely to become more separated from farm household activities, one may expect this bias to be increasing over time.

A yet more important shortcoming of Mundle's estimates is that he calculates agricultural purchases and sales at purchasers' prices. While this is the correct procedure as far as agricultural purchases are concerned, it introduces an important overestimation of agricultural sales, which cannot be ignored. According to this procedure, the income of households involved in the transportation and trade of agricultural goods is included in the income of agricultural producers. This is a serious shortcoming, which has to be overcome by subtracting the transport and traders' margins from the value of agricultural sales. According to Karshenas (1990*b*), this margin accounted for 23–25 per cent of agricultural sales in Iran. Given that India is a much larger country, with a higher degree of specialization in agricultural production in different regions, and possibly a lower degree of efficiency in transport and trade, the margin may be much higher here. We have assumed a 25 per cent traders' margins in recalculating Mundle's estimates. This is shown as the 'new estimates' included in Table 5.1 and Figure 5.1, alongside Mundle's original estimates.

As can be seen from Table 5.1 and Figure 5.1, Mundle's estimates indicate a net

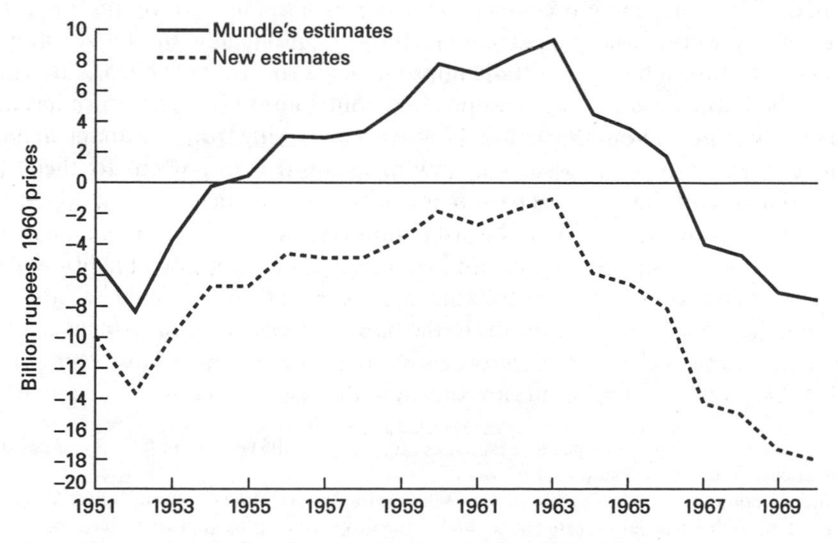

FIG. 5.1 Net surplus outflow from agriculture, India, 1951–1970

inflow of resources into the agricultural sector in the early 1950s, turning into an outflow from the mid-1950s and a relatively large drain of resources from agriculture by the early 1960s. This trend is again reversed from the mid-1960s, and by the end of the decade there is a relatively large inflow to agriculture. By contrast, the 'new estimates' show a negative financial contribution of agriculture throughout the period. It should be noted, however, that the new estimates differ from Mundle's estimates only in absolute value rather than in trends over time (Figure 5.1). As we noted in Chapter 2, the absolute value of real net intersectoral resource flows very much depends on the base year or reference point prices, and, with the adoption of a different base year, the new estimates may also indicate a resource outflow from agriculture in some of the intervening years. This is very likely, as the terms of trade shown in Table 5.2 indicate a substantial improvement in the agricultural terms of trade, particularly in the second half of the 1970s. We shall return to this point shortly. Given how sensitive the estimates of real net intersectoral resource flows are likely to be to the choice of base year, it may be more informative if we compare the new estimates with those of Mundle at current prices. Current price estimates of net intersectoral resource flows are shown in the first two columns of Table 5.3. As can be seen, while the new estimates indicate a net inflow of financial resources to agriculture throughout the period, Mundle's estimates, with the exception of the first few years, indicate a financial outflow throughout the period. The new estimates seem to be supported by independent estimates of the net resource transfers from the financial side, all of which show a net resource transfer into agriculture on both the private and government current and capital transfer accounts[3] (see e.g. Krishna and Raychaudhuri, 1980; Mody, 1981; Shetty, 1971; Mody, Mundle, and Raj, 1985).[4]

Indian agriculture therefore seems to have been a financial drain on the rest of the economy during the period under study. Combined with the substantial income gains through terms of trade improvements shown in Table 5.2, this may convey the impression that the real product contribution of agriculture has also been negative throughout the period. However, in moving from the financial flows at current prices to the real surplus flow measure, it is important to check the appropriateness of the assumed base year prices. The income gains from the terms of trade improvements shown in the last column of Table 5.2 are in fact measured at 1951 base year prices. But, as we saw in Chapter 2, the adoption of a different set of base year prices may put the results at variance from those depicted in Table 5.2. To check the sensitivity of the results to the base year prices, we have measured the new estimates of real net intersectoral resource flows at three base year prices, 1951, 1958, and 1968. The results are shown in the last three columns of Table 5.3.

[3] The only item for which independent estimates do not exist is the net flow of factor incomes. This will be discussed further in Section 5.4 below.

[4] After careful review of the existing studies, Mody, Mundle, and Raj (1985) conclude: 'Whichever source we use, and whatever adjustment we make, the farm sector thus appears to have been a net borrower.'

Table 5.2 Terms of trade of agriculture, India, 1951–1970

Year	Price indices		Net barter terms of trade (P_a/P_n)	Terms of trade in income[a]
	Agricultural sales (P_a)	Agricultural purchases (P_n)		
1951	100.00	100.00	100.00	0.00
1952	86.40	87.78	98.43	−0.49
1953	91.00	88.35	103.00	0.85
1954	82.63	85.77	96.33	−1.02
1955	77.36	82.21	94.10	−1.91
1956	90.07	88.54	101.73	0.47
1957	92.19	94.31	97.75	−0.65
1958	97.12	96.22	100.93	0.28
1959	100.44	99.49	100.95	0.28
1960	104.78	105.53	99.29	−0.22
1961	105.21	105.23	99.97	−0.01
1962	106.83	108.59	98.38	−0.52
1963	113.42	117.30	96.70	−1.09
1964	132.90	123.18	107.89	2.69
1965	144.78	127.38	113.66	4.45
1966	168.83	138.17	122.19	7.04
1967	187.15	150.77	124.13	8.99
1968	183.20	158.70	115.44	6.31
1969	198.85	159.30	124.83	10.01
1970	208.07	164.59	126.41	10.81

[a] bn. rupees, at 1951 prices, calculated as $m_a\ (1 - P_n/P_a)$, where ma is the real value of purchases of the agricultural sector.

Source: Mundle (1981).

As can be seen, the two base years 1951 and 1958 show similar results which indicate an inflow of real net resource flows into agriculture throughout the 1951–70 period. In fact, since the agricultural terms of trade during the 1951–63 period remained more or less stable, any base year from this period would be expected to yield very close results. However, the choice of a base year from 1964 onwards, when agricultural terms of trade show a steep increase, affects the results more significantly. For example, as shown in Table 5.3, using 1968 base year prices, the net surplus flow measure shows an outflow of resources from the agricultural sector for the four years between 1959 and 1963 – though the outflow remains relatively small, and, indeed, by 1970 it turns into a substantial inflow similar to the earlier two measures.

The question is: which of these three years, if any of them, comes closest to what may be regarded as an appropriate base year for the measurement of the terms of trade effect? To answer this question it is important to note that, throughout the period under study, the Indian government constantly intervened in the market to control food and other agricultural raw material prices – though the nature and the

Table 5.3 Resource outflow from Indian agriculture at current and constant prices, 1951–1970

Year	At current prices		Real values at base year prices[b]		
	New estimates	Mundle's estimates[a]	1951	1958	1968
1955	−9.86	−4.49	−9.86	−9.34	−11.70
1952	−12.17	−7.16	−13.59	−12.92	−17.31
1953	−8.15	−2.25	−9.81	−9.26	−10.80
1954	−6.40	−0.85	−6.73	−6.29	−5.73
1955	−6.59	−0.87	−6.71	−6.26	−5.22
1956	−3.73	3.27	−4.62	−4.23	−1.61
1957	−5.01	2.23	−4.78	−4.39	−1.82
1958	−4.46	3.63	−4.87	−4.46	−1.61
1959	−3.36	5.38	−3.63	−3.26	0.64
1960	−2.05	7.90	−1.74	−1.42	4.22
1961	−2.73	7.56	−2.59	−2.22	3.09
1962	−2.27	8.35	−1.61	−1.28	4.76
1963	−2.30	9.41	−0.94	−0.63	6.10
1964	−4.00	9.76	−5.70	−5.21	−1.44
1965	−2.84	11.94	−6.41	−5.90	−2.67
1966	−1.83	15.41	−8.12	−7.54	−5.38
1967	−9.57	10.48	−14.10	−13.28	−14.51
1968	−15.62	4.14	−14.84	−13.98	−15.62
1969	−14.54	7.33	−17.32	−16.37	−19.40
1970	−15.01	8.37	−18.02	−17.04	−20.35

[a] Evaluated on the bases of Mundle's estimates of real sales and purchases in Table 5.1 and the terms of trade data in Table 5.2.
[b] New estimates.

Source: Based on Tables 5.1 and 5.2.

purpose of intervention differed during different subperiods.[5] According to Mundle (1981), three broad subperiods could be distinguished. Up to the mid-1950s government intervention was mainly directed towards moderating annual fluctuations in food prices, and it did not exert any significant influence on price trends. During the latter half of the 1950s and the first half of the 1960s, however, government intervention was directed more towards moderating the trend increase in food prices, and hence increasing resort was made to food grain imports.[6] During the third subperiod, that is, from 1964 onwards, the government adopted a

[5] For a detailed discussion of the agricultural price policy in this period, see Mundle (1981, ch. 6).

[6] Given the trade control regime in force during this period, the mere fact that imports were growing, or that the government purchase price for grain was below the prevailing market price, does not necessarily indicate a price-moderating influence by the government. As we shall observe shortly, agricultural terms of trade in India relative to those of the world market showed an improvement in favour of agriculture throughout this period. However, from the point of view of the attitude of government towards agricultural prices, this period is distinct from the other two. It is in this sense that it is considered to be marked by a policy of low agricultural prices, particularly compared to the latter half of the 1960s.

high agricultural price policy as part of the new agricultural strategy, and, as we have already seen, this period witnessed a steep terms of trade improvement in favour of agriculture.

Judging by what Ishikawa defines as a 'normal' year (see Chapter 2), surely the period 1964–70 does not qualify as an appropriate choice for the base year. Given the time-lags involved in non-agricultural prices adjusting to agricultural price changes, and the even longer time-lags for quantities to adjust, a period of steep agricultural price increases such as this can hardly be said to be close to an equilibrium situation. Any of the years during the earlier periods, particularly those of the early to mid-1950s, when the terms of trade remained relatively stable and government intervention did not involve forcing the price trends in any particular direction, may make a better 'normal' year *à la* Ishikawa. On this basis, the estimates with 1951 as the base year may be said to be more accurate indicators of the direction of real surplus transfer than those using 1968. Judging on the basis of world prices, however, a different picture might emerge.

As was noted above, throughout the period under study both the agricultural and non-agricultural traded goods were subject to trade restrictions – though admittedly the non-agricultural consumer goods enjoyed a higher degree of protection. On this basis, the relative prices prevailing in 1968 may be said to come closer to world prices. No study has been made of the degree of protection afforded to the output and inputs of the agricultural sector over this period. But it would be illuminating to compare the trends in the agricultural terms of trade in the Indian economy with those in the world market. The indices of the Indian and world terms of trade are shown in Figure 5.2. It is interesting to note that there was a continuous improvement in agricultural terms of trade in India throughout the period, and this continued even during the 1956–63 period, when it is said that government intervention was directed towards depressing agricultural prices.[7] By 1970 the agricultural terms of trade in India were more than 90 per cent higher than their international counterpart, measured against their respective levels in 1951. In other words, had the terms of trade in India followed the same path as world prices, by the end of the period the income loss to the Indian farmers would have been substantial.[8] Under these circumstances, whether the earlier years or the later years come closer to approximating world prices depends on the relative protection given to the output and inputs of the agricultural sector in 1951. As mentioned above, we have little information on this, but perhaps it is not unreasonable to assume that the difference was not as high as the 90 per cent wedge which

[7] The world terms of trade series is based on the data provided by Grilli and Yang (1988), who obtain manufacturing prices from the manufactured export prices of industrialized countries. For agricultural prices, we have calculated a composite index of the food and non-food agricultural prices based on Grilli and Yang's data, using a 2:1 weighting system which is equal to the average ratio for sales of consumer and producer goods by Indian agriculture over the study period.

[8] More precisely, the income loss by 1970 would have been equal to Rs 27bn. at 1951 prices, compared to the actual income gain of about Rs 11bn. shown in Table 5.2.

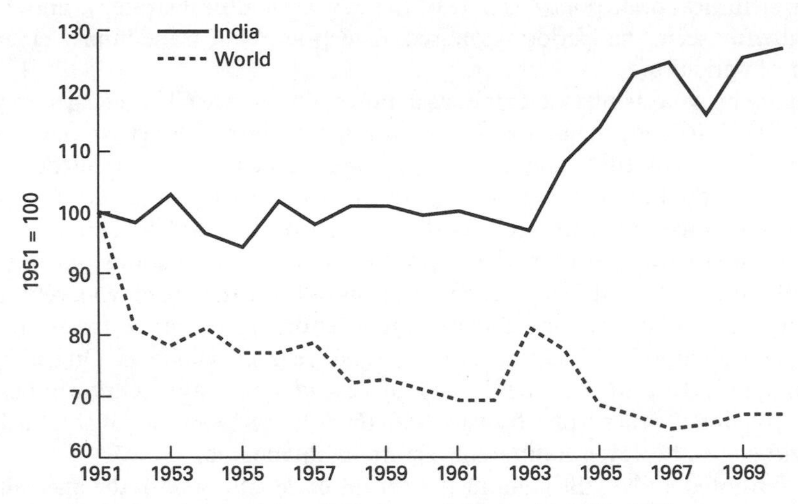

FIG. 5.2 Agricultural terms of trade, India and the world, 1951–1970

developed between the agricultural terms of trade of India and the world level by the end of the study period.[9] On the basis of this conjecture, it could be argued that a base year around the mid-1960s, perhaps 1968, comes reasonably close to approximating world relative prices.

The above clearly shows the problems associated with the search for an absolute measure of real surplus transfer, as discussed in Chapter 2. As we mentioned there, the pattern and the change in surplus transfer in the course of economic development are often more informative than the absolute level of real net resource flows. These issues are the ones on which we shall concentrate in the remaining part of this chapter. However, it is important at this stage to recapitulate on the conclusions that we can reach on the direction of net resource flows from the above discussion. As far as the net intersectoral resource flows at current prices are concerned, the existing evidence seems to suggest a net inflow of financial resources into the agricultural sector in India throughout the period under study. In real terms, also, the real net financial contribution of agriculture seems to have been negative throughout the period, measured at the base year prices prevailing between 1951 and 1965. Using the last few years of the 1960s as the base year, however, real net surplus transfer from agriculture becomes positive for a number

[9] Another way of looking at this problem is to check the extent to which the terms of trade in India at the beginning of our study period diverged from the world prices in a historical context. According to Tyagi (1979, table 14), there was about a 26% increase in agricultural terms of trade in India between 1939 and 1947. According to the data provided by Grilli and Yang (1988), agricultural terms of trade at the world level increased by about 31% over the same period. The 5% wedge between Indian and world relative prices in the earlier period clearly does not justify the 90% change in the opposite direction during our study period.

of years, though it still shows a substantial inflow of resources into agriculture during much of the initial and final period under study. This cyclical behaviour in movement of surplus transfer, which is visible in all the different estimates, whether at current or constant prices, illuminates some of the important features of the development of Indian agriculture which will be discussed below.

5.3 Determinants of Resource Flows

Setting aside the question of the absolute value and the direction of net resource flows, Mundle's estimates show important aspects of the changing *trends* in intersectoral resource flows over the 1951–71 period which are also replicated in our new measure at both current and constant prices. Disregarding the cyclical movement of net resource flow in the early 1950s,[10] one can observe a secular trend of declining resource inflow into agriculture which, from the mid-1960s, is reversed into a growing magnitude of resource inflow. To distinguish more clearly the factors responsible for this reversal of trends, we have shown the growth rates in real terms in intersectoral commodity flows over the 1955–64 and 1964–70 sub-periods in Table 5.4. As can be seen, the deceleration in the growth of agricultural sales and the acceleration in the growth of agricultural purchases are equally responsible for the reversal of the intersectoral net commodity flow trends. These results may appear paradoxical at first, particularly in view of the fact that there seems to have been a slight decline in the rate of growth of agricultural output in the latter period.[11]

The changing trends of intersectoral commodity flows are the result of complex forces on both the demand side and the supply side. Given the more or less elastic

Table 5.4 Growth of intersectoral commodity exchanges, India 1955–1977

	Real average annual growth rates	
	1955/6–1964/5	1964/5–1970/1
Agriculture's imports	*2.7*	*5.8*
Consumer goods	2.5	4.4
Producer goods	4.4	13.5
Agriculture's exports	*3.8*	*1.3*
Consumer goods	2.1	2.9
Producer goods	6.6	−1.1

Source: Table 5.1.

[10] As pointed out by Mundle (1981), the estimates for the first three years are based in part on different data definitions, and thus may not be comparable to the estimates for the rest of the period.

[11] The annual average rate of growth of real gross product in agriculture declined from 2.9% during 1951–64 to 2.4% during 1964–70. Though there has been some controversy as to whether the agricultural growth rate actually declined, there is no claim that it increased (see Shetty, 1978; Srinivasan, 1979; Mundle, Mody, and Raj, 1985).

supply of industrial goods, it could safely be assumed that the agricultural purchases are largely demand-determined. The acceleration in the growth of imports of producer goods into agriculture in the latter period largely reflected the spread of the new biochemical technology in agriculture with the inception of the green revolution period from the mid-1960s. This seems to have led to a substantial increase in the demand for non-agricultural inputs per unit of agricultural output.[12] However, as producer goods had a very small share of total purchases in the earlier years, consumer goods purchases played a more important role in the acceleration of the growth of total imports into the agricultural sector from the mid-1960s.

The most important factor in explaining the growth of real consumption in agriculture is the growth of real income of the farm households. Assuming the non-agricultural consumption goods to have a higher income elasticity of demand, it would also be plausible to assume that, the more skewed the distribution of the new incomes generated in the sector, the higher the increase in the non-agricultural component of the increment in farm household consumption. Real income of the agricultural households grows principally as a result of the normal growth of real value added in the agricultural sector and/or the growth of the purchasing power of that income through the improvements in the terms of trade. Paradoxically, total agricultural value added seems to have a slightly lower growth rate in real terms in the latter period. In explaining the acceleration of the growth of consumer goods imports into agriculture, however, as noted in Chapter 3, it is the growth of output per agricultural worker which is more relevant. The available data on employment in the agricultural sector during the period under study are somewhat contradictory. On the one hand, considering the data on agricultural population, we can observe an acceleration in the rate of growth during the 1964–70 period compared to the 1951–64 period.[13] On this basis, the data suggest a clear deceleration in the rate of growth of value added per head of the agricultural population in the latter period. On the other hand, the available labour statistics suggest a noticeable deceleration in the rate of growth of employment in the agricultural sector during the 1960s.[14] According to this evidence, output per agricultural worker during the later 1960s seems to have had a higher rate of growth than in the 1950s and the early 1960s. While agricultural output per worker declined by about 31 rupees (in 1970 prices) over the 1950–61 period, it increased by about 25 rupees over the 1961–71 period (Chakravarty, 1987, p. 113). The accuracy of the official statistics

[12] The trend rate of increase in input/output ratio of nonagricultural inputs in agricultural production rose from about 2% between 1955 and 1965 to about 11% over the 1965–70 period (calculated from Mundle, 1988, table 4.7).

[13] The rate of growth of the agricultural population seems to have increased from about 1.9% per annum during the 1951–64 period to 2.4% during the 1964–70 period (Mundle, 1981, table 3.7, p. 56).

[14] According to the data in Raghavan (1984), also reported in Chakravarty (1987), the number of agricultural workers increased by 3.06% per annum during the 1950–60 period and by 1.96% per annum during the 1960–70 period.

on employment in the agricultural sector has been questioned by some Indian economists (see e.g. Chaudhuri, 1978, pp. 42–9; Krishnamurthy, 1974). However, even if the labour statistics were accurate, the implied increase in labour productivity is not significant enough to explain the rapid acceleration in the rate of growth of demand for non-agricultural consumer goods in the agricultural sector. A more important explanation is the income gains in agriculture arising from terms of trade improvements.

As pointed out above, there was a rapid improvement in agricultural terms of trade from the mid-1960s, leading to a substantial income gain for the agricultural sector (see Table 5.2).[15] To realize the significance of this factor, one need only point out that the income gains in agriculture from terms of trade improvement between 1964 and 1970 were Rupees 8.1bn. (in 1951 prices), which compares with a Rs, 4,5bn. increase in real agricultural value added (in 1951 prices) over the same period. In other words, income gains from terms of trade changes were more than twice those from normal growth of agricultural output. This would surely have had a considerable effect on the growth of consumption. Furthermore, the income gains resulting from the terms of trade improvements are likely primarily to have benefited the richer sections of the agricultural households, namely, those who produce the major part of the marketed surplus. One would thus expect a faster growth of consumption of non-agricultural consumer goods with higher income elasticity of demand relative to the total consumption of agricultural households – evidenced by the rapid acceleration in the growth of purchases by agriculture from other sectors for final consumption.

As regards the decline in the rate of growth of agricultural sales from the mid-1960s, the explanation may be sought in various demand- and supply-side factors. Mundle, Mody, and Raj (1985) have emphasized the demand-side factors. According to them, the slower growth of consumer goods sales is explained partly by the slower growth of the non-agricultural sector, and partly by the negative income and price effects of the shift in terms of trade from the mid-1960s. In the case of producer goods sales, they point out the additional factor of relative decline in the share of agro-based industries in the industrial output (ibid., p. 283). Mundle (1981) also mentions structural changes in the non-agricultural sector which led to a decline in the use of agricultural inputs per unit of non-agricultural output during the 1960s. These explanations are certainly correct in identifying the proximate or immediate causes of this phenomenon, and can be regarded as adequate if we assume that the slow-down in non-agricultural growth from the mid-1960s was an autonomous phenomenon independent of what was happening in agriculture. The

[15] There is a vast literature and some controversy on the terms of trade movement during the period under study (see e.g. Mitra, 1977; Tyagi, 1979, 1988; Kahlon and Tyagi, 1980; Vittal, 1986, 1988; Ghosh, 1988). In this study we have used estimates by Thamarajakshi (1969), as these are the most consistent time-series estimates available and have also been used by Mundle (1981), which is the only available source on intersectoral commodity flows.

demand-side explanations, however, do not seem entirely satisfactory, given that there was a sharp acceleration in the imports of agricultural products from the rest of the world over this period,[16] and given the agricultural price increases since the mid-1960s. This may be indicative of the fact that the ultimate cause of both the slow-down in industrial growth and the deceleration in agricultural sales was the slow growth of productivity and output in the agricultural sector during the second half of the 1960s. Other independent studies have also emphasized the food-supply constraint to industrial growth in India over this period (see e.g. Sen, 1981; Chaudhuri, 1987). To this, however, should be added the massive income gains in agriculture due to the terms of trade effect, which, *ceteris paribus* and assuming food to be a normal good, would be expected to have led to a decline in marketed surplus of the agricultural sector in the 1960s.[17]

In analysing the process of net surplus transfer between agriculture and the rest of the economy, it appears that the distinction between income gains which accrue as a result of the normal growth of real value added in agriculture and those which accrue as a result of terms of trade improvements is a crucial one. The former increases the demand for non-agricultural consumer goods in the agricultural sector, and finances this demand by increased marketed surplus. The income gains which accrue purely as a result of the terms of trade improvements, on the other hand, increase the demand for consumption goods imports into the farm sector without a concomitant rise in marketed surplus, perhaps even accompanied by a contraction in marketed surplus. The net result is likely to be a double contraction in the net outflow of resources from the agricultural sector on the consumption accounts. Of course, the terms of trade changes are only one of the mechanisms of income transfer between sectors on the financial side. Income gains through terms of trade improvements may be siphoned off through increased taxation by the government, or they can be moved out as a result of voluntary savings by the farm households through the credit system. This leads us to the financial side of the resource flow process, which is the subject-matter of the next section.

[16] Net agricultural imports from the rest of the world, which were less than Rupees 1bn. (in 1960 prices) for the entire 1950–64 period, increased to Rs 6.82bn. (1960 prices) over the 1964–70 period; that is, fifteen-fold on an average annual basis.

[17] The strength of this effect, of course, depends on the stratification of the farm households and control over marketed surplus by different farm income groups. According to estimates by Patnaik (1975), 70% of the farm households with holdings of less than 5 acres controlled only about 18% per cent of the operated land area and provided no more than 16% of the total marketed surplus. The income gains through the terms of trade effect are thus expected to have been concentrated in the moderately well-off to rich peasant households with relatively lower income elasticity of demand for food. To this extent, the contraction in marketed surplus would have been moderated. I am grateful to Utsa Patnaik for bringing this point to my attention. For earlier studies of the relation between farm household stratification and agricultural marketed surplus, see Narain (1961) and Sanghvi (1969).

5.4 Financial Decomposition of Surplus Flow

The intersectoral commodity flows have their counterpart in the financial flows, as discussed in Chapter 2. Table 5.5 shows the different financial flows which add up to the balance on commodity exchanges at current prices. These should be regarded as tentative estimates based on partial studies by different authors.[18] Since no independent study of net factor payments is available, this item is estimated as a residual, which is reported both on the basis of Mundle's estimates of net commodity flows and the 'new estimates'.

All the independent estimates indicate a net financial inflow into agriculture on both the private and official accounts.[19] For Mundle's estimates to be correct, we need a relatively large countervailing net factor income outflow ($F_a - Y_f$) equivalent to 7 per cent of the agricultural value added, which does not seem to be plausible. On the other hand, the new estimates indicate a net factor income inflow of about 5 per cent of total agricultural income;[20] in other words, labour income plus net private current transfers are greater than the value of rent to non-farming landlords and interest payments.[21] Though inter-country comparisons with regard to net factor income flows can be problematic, it may still be pointed out that these latter estimates are much closer to the experience of China, Japan, Iran, and post-colonial Taiwan, discussed in later chapters of the book.

As for the savings surplus of the agricultural sector (AS or K, as defined in Equations 2.10 and 2.11), the figures in row 4 of Table 5.5 indicate that the contribution of agriculture in this regard has also been negative. Estimates by Ishikawa (1967*a*), Mody (1981), and Shetty (1971) suggest that, in fact, net surplus flow from agriculture has been negative throughout the 1951–70 period, both on the private and government side of the accounts.[22] Due care, however, must be

[18] Note also that there is a problem in synchronizing the dates for the different items reported in the table. That is, while official current and capital transfers based on Shetty (1971) relate to the average for the years 1966–8, the rest of the items refer to 1968.

[19] The data provided for 1968 in Table 5.5 are valid for other years during the study period as well. This was the case for both capital and current transfer accounts in all the independent studies of different years during the 1951–70 period by different authors (see e.g. Ishikawa 1967*a*; Mody, 1981; Shetty, 1971).

[20] Note that the new estimates are based on a trade and transport margin of 25%. If we alternatively assume a 20% margin for trade and transport margin in agricultural sales we still get a factor income inflow of about 1% of agricultural value added (see Karshenas, 1989).

[21] On the basis of the findings of other independent studies, this seems to be a more reasonable condition than that implied by Mundle's estimates. Ishikawa (1967*a*) found that, in 1951, there was a net factor income inflow into the farm sector which constituted about 8% of net resource inflow into the sector. According to Mody (1981), there is no evidence to suggest 'that since the early 1950s the direction of net factor income flows has been reversed'. Other indirect evidence also seems to support this view. On the one hand, with the abolition of the *zamindari* system, India did not have a large absentee landlord class during this period. On the other hand, factor income through wage labour in non-agricultural activities seems to have formed a relatively large source of income for Indian agricultural households (see e.g. Mellor *et al.*, 1968, pp. 309–29).

[22] For a review of these different estimates, see Mody *et al.*, 1985.

Table 5.5 Financing of net agricultural resource outflow, India, 1968/9 (bn. rupees, current prices)

	Mundle's estimates		New estimates	
Net resource outflow (R)	4.14	(3)	−15.62	(−10)
(*a*) Agriculture's sales (X)	79.03	(50)	59.27	(37)
(*b*) Agriculture's purchases (M)	74.89	(47)	74.89	(47)
Financing items				
1. Net outflow of factor income[a]($F_a - Y_f$)	11.35	(7)	−8.41	(−5)
2. Net outflow of current transfers ($T_{fg} - T_{gf}$)	3.95	(2)	3.95	(2)
3. Net outflow of capital transfers	−11.16	(−7)	−11.16	(−7)
(*a*) Private ($K_{fo} - K_{of}$)	−6.46	(−4)	−6.46	(−4)
(*b*) Public[c]($K_{fg} - K_{gf}$)	−4.70	(−3)	−4.70	(−3)

Note: figures in brackets are % share of agricultural income.

[a] Includes current transfer of private sector.

[b] Refers to government's tax receipts only; current official inflows included in 4 (*b*).

[c] Includes government current expenditure; refers to average for 1966–8.

Source: Mundle (1981), Mody, Mundle, and Raj (1985).

taken in interpreting these results. For example, as pointed out in Table 5.5, public sector capital flows (row 4(*b*)) also include current government expenditure in the farm sector. On the other hand estimates of private capital flows by Mody (1981)

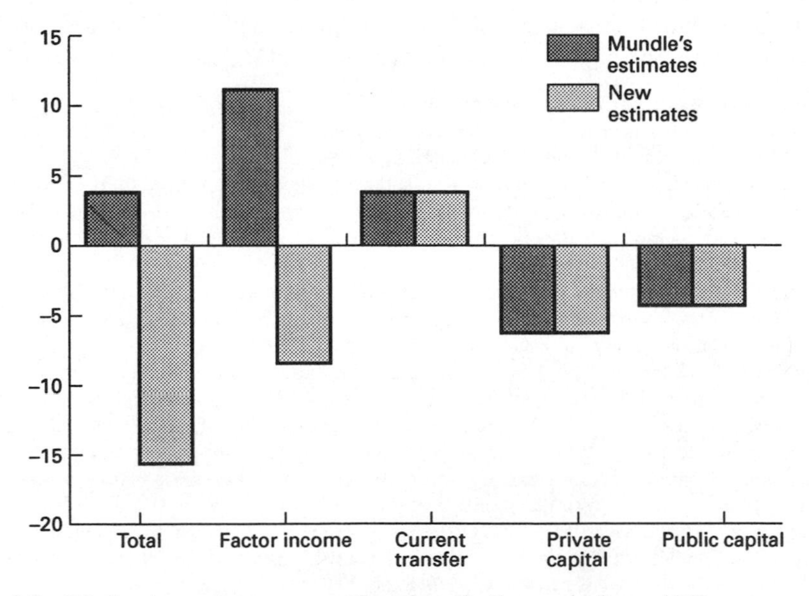

FIG. 5.3 Mechanisms of resource outflow from Indian agriculture, 1968

exclude the capital flow through informal credit markets.[23] To the extent that this flow is composed of credits and loans to suppliers by traders or absentee landlords, it constitutes an inflow into the farm sector, and hence should be included in the estimates. The inclusion of current government transfers would thus tend to inflate the estimates of net capital inflow, while the exclusion of informal credit inflow leads to an underestimation. On balance, these amendments are unlikely to change the direction of capital flows; if anything, they would perhaps increase the size of capital inflow. One can thus safely conclude that, as with the net product or net finance contribution of agriculture, the net savings surplus also was negative in the period under study.

As with the analysis of surplus transfer from the real side, it would be instructive to consider the behaviour of different mechanisms of financial transfer over time and in a dynamic context. In particular, a comparison between the fourteen years before the new agricultural strategy and the years in the latter half of the 1960s would be illuminating. We have already discussed one mechanism of financial transfer, namely, the terms of trade effect, in the previous section. The question now is to what extent the other financial mechanisms have complemented the income gains through terms of trade effect in the aftermath of the new agricultural strategy, and to what extent they have acted as a mechanism for siphoning off this income. The estimates of taxable capacity and tax burden of the farm and non-farm sectors by Shetty (1971) indicate that, in fact, the latter half of the 1960s witnessed a decline in the tax/income ratio in the farm sector, despite the large increases in the taxable capacity in the sector. Taxable capacity of the farm sector – as determined by factors such as the level and distribution of income and a measure of the minimum consumption requirements and the related investment needs – increased from about 23 per cent of the sector's income in the first half of the 1950s to about 28 per cent in the early 1960s and 46 per cent in the 1966–8 period, according to Shetty's estimates. On the other hand, the actual tax burden, as measured by the tax/income ratio, which grew from 5.1 per cent to 9.1 per cent during the 1951–65 period, actually declined to 7.9 per cent in the latter half of the 1960s.[24] In other words, the actual to potential tax ratio seems to have experienced a 20 per cent decline between the early 1950s and the late 1960s, most of which was concentrated in the latter half of the 1960s.

Whether measured in terms of actual tax burden or potential taxable income, clearly the Indian government seems to have failed to skim off the income gains in the farm sector, particularly in the aftermath of the adoption of the new agricul-

[23] It should be noted that the major part of the Rs 6.46bn. net private capital flow in Table 5.5 is from government banks and government-financed co-operatives. However, as mentioned in Ch. 2, we have followed the convention of regarding all bank lendings as private capital flows.

[24] The major part of the decline was concentrated in direct tax ratio, which declined by 44%, from 1.6% in 1961–5 to 0.7% in 1966–8. Indirect tax ratio over the same period declined by only about 8%, from 7.6 to 7.0% (Shetty, 1971).

tural strategy.[25] Given the lack of a large absentee landlord class which could act as a significant source of surplus transfer from agriculture, the only other financial channel for surplus flow was through voluntary savings and asset acquisition of the farm households. As we have already seen, however, according to Mody's estimates, the farm households seem to have been net borrowers throughout the period under study. In other words, the spending of the farm sector was well above its disposable income and the gap between the two, as we observed in the previous section, widened even further from the mid-1960s. This was partly due to the increased spending on new producer goods with the green revolution era, but, even more, to the increase in consumer spending.

5.5 Concluding Remarks

The empirical evidence discussed in this chapter suggests that the net finance contribution of the agricultural sector in India during the 1950–71 period was negative. There seems to have been a net resource inflow into Indian agriculture during this time. This is brought out by independent estimates on both the real side, i.e. the intersectoral flow of commodities, and the financial side, namely, the private and public current and capital transfers and the terms of trade effect. We have also observed the changing trend of net resource transfer in different stages of agricultural growth. Up to the mid-1960s, there was an increasing trend in the rate of marketed surplus, while purchases of the agricultural sector as a proportion of agricultural value added remained more of less stable, thus leading to a narrowing gap in the net surplus inflow into agriculture over time. With the new agricultural strategy from the mid-1960s, however, these trends seem to have sharply reversed, leading to a rapid increase in the net inflow of resources into agriculture.[26]

The new agricultural strategy was introduced at a time when traditional agriculture seemed to be faltering as a result of the virtual exhaustion of new cultivable land frontiers and growing population pressure on land. Technological modernization of the sector was perceived as a necessary precondition for the resumption of agricultural growth. Central to the new strategy was the introduction of the land-saving technology of the green revolution. This involved a shift in the input base of the agricultural sector towards increasing use of fertilizers, new varieties of fertilizer-intensive seeds, and other related inputs, notably water and commercial energy sources such as electricity and oil. Other elements of the new strategy

[25] Other estimates provided by Gandhi (1966) also show that, in the period 1950–60, tax revenues from agriculture were substantially below government expenditure in agriculture. In addition, he shows that the tax rates of different income brackets in rural areas were considerably lower than their urban counterparts – particularly in higher income groups.

[26] According to estimates by Mundle (1981), the ratio of marketed surplus to agricultural value added increased from an average of 43.7% during 1951–3 to 60.2% during 1963–5. The ratio of purchases of the agricultural sector to value added during the same period declined from about 53% to 52%. By 1970, however, the marketed surplus ratio had declined to 52.5%, and the purchase ratio had increased to 61.3%.

consisted of increased provision of credit and some shift in emphasis from publicly financed large irrigation projects to small tube-wells and energized pump sets (Chakravarty, 1987). There was also a change in the terms of trade of the agricultural sector, through government price support policies and input subsidies. A direct impact of the new policy on intersectoral resource flows was the rapid growth of agricultural purchases of producer goods from other sectors. The effect on the net resource flows crucially depended on the resulting productivity growth in agriculture, and also on the way in which the new income generated in the sector as a result of this productivity growth was utilized.

The new agricultural policy seems to have had limited success in improving agricultural performance. In comparing the trends of the decade following the new strategy with those of the fifteen years preceding it, Srinivasan (1979) sums up the results of his detailed statistical study as follows: 'While there has been a decline in the rate of growth of gross sown areas, in particular under non-food crops in the decade starting from 1967–68 compared to the fifteen years ending in 1964–65, the output (and yield per unit area) of food crops and all crops grew more or less uniformly over the entire period with no evidence of either acceleration or deceleration since 1967–68.' The impact of the new strategy was restricted to a few crops such as bajra, maize, and particularly wheat, which achieved impressive growth rates. With regard to most other crops, one can in fact detect a deceleration in the rates of growth during the green revolution period. The impact of the new strategy also seems to have been mainly restricted to only a few states which were already well endowed in terms of the infrastructure, irrigation, and agrarian institutions necessary for the effective deployment of the new technology.[27]

The restricted response of agricultural output to the increased use of new inputs explains the rapid rise in the ratio of producer goods purchases to agricultural value added from the mid-1960s. However, to explain the other items which comprised the net intersectoral surplus flows, notably the flow of consumer goods, one needs also to consider the financial side. As we have already noted, the agricultural sector in India made considerable income gains from favourable terms of trade movements, particularly in the latter half of the 1960s, when the new technology was introduced. Another notable feature of the financial flows was the extremely low tax burden on the agricultural incomes relative to the incomes generated in other sectors of the economy, and particularly the decline in the tax burden in the latter half of the 1960s. In the absence of other major sources of financial outflow from agriculture, the income gains in the farm sector were translated into increased spending, particularly consumer spending. This seems to have led to a double squeeze on surplus outflow from agriculture; first, by increasing the farm sector's purchases of consumer goods from other sectors, and, secondly, by increasing own consumption of farm producers, and hence reducing

[27] There is a vast literature on this topic which cannot be discussed here in any detail. For some useful sources and further references, see Narain (1976), Srinivasan (1979), Shetty (1978), Sen (1981), Chaudhury (1978).

the marketed surplus ratio. In many ways, the experience of India with regard to intersectoral resource flows stands in sharp contrast to that of Taiwan, which will be discussed in the next chapter.

6

Intersectoral Resource Flows in Taiwan, 1911–1960

6.1 Introduction

In this chapter we examine the intersectoral resource processes in Taiwan during the period 1911–60. A detailed analysis of intersectoral commodity flows and capital transfers for Taiwan over this period is provided by Lee (1971).[1] These estimates, which stretch over the pre- and post-colonial periods, reveal a number of interesting features of changing intersectoral relations in the process of development. In the next section we discuss the overall patterns of intersectoral resource flows and their constituent elements. It will be shown that, throughout the period under study, there was a considerable outflow of resources from the agricultural sector. The rapid pace of productivity growth in agriculture provided the conditions for the possibility of resource extraction from the sector on a sustained basis. Section 6.3 looks at the development of the agricultural sector, which, in a sense, provides some of the key determinants of surplus flow on the real side. Section 6.4 discusses the different financial mechanisms of surplus transfer, including the terms of trade effect, and a summary and the conclusions of the chapter are given in Section 6.5.

6.2 Resource Flows on the Real Side

The flow of real commodity exchanges between agriculture and non-agriculture for the 1911–60 period is given in Table 6.1. A striking feature of Taiwan's experience is the extremely large resource outflows from the agricultural sector throughout this time. Net resource outflow from the agricultural sector constituted about 40–50 per cent of gross sales of the sector during the 1910s and the 1920s, declined to about 30 per cent in the 1930s, and remained more or less at this ratio for the rest of the period. As noted in the table, these estimates are made on the basis of the reference point or base year prices of the 1935–7 period. The change in the base year prices certainly changes the magnitude of the estimated net resource flows. However, given the substantial size of the resource outflow and the

[1] The study by Lee (1971) in fact covers the period 1895–1960. However, since various components of resource flows for the earlier years are missing, we focus only on the 1911–60 period.

Table 6.1 Real intersectoral commodity flows, Taiwan, 1911–1960 ($Tm., 1935–7 prices)

Year	Real value of agricultural exports			Real value of Agricultural imports			Net balance (X–M)
	Consumer goods	Producer goods[a]	Total (X)	Consumer goods	Producer goods	Total (M)	
1911–15	30.0	61.7	91.7	32.9	9.6	42.5	49.2
1916–20	27.6	96.5	124.1	37.0	15.9	52.9	71.2
1921–5	40.2	111.8	152.0	59.6	32.5	92.1	59.9
1926–30	46.6	151.4	198.0	84.5	54.3	138.8	59.2
1931–5	60.0	200.0	260.0	105.8	64.0	169.8	90.2
1936–40	68.3	233.4	301.7	137.4	74.8	212.2	89.5
1950–5	138.2	159.6	297.8	119.4	65.6	185.0	112.8
1956–60	168.1	221.0	389.1	165.6	127.4	293.0	96.1

[a] Sales of producer goods includes overseas exports.

Source: Lee (1971).

relatively small share of the terms of trade effect in the total flow, the results are not significantly affected by the change in the base year,[2] and certainly the trends in real net resource flows over time remain the same. We shall therefore be considering the real flows at 1935–7 prices throughout this chapter.

Both the sales and purchases of the agricultural sector were growing relatively fast during the period under study. Sales, starting at a higher level than purchases, also achieved a higher rate of growth, so that up to the mid-1950s the gap between the two was widening. This trend seems to have been partially reversed during the 1956–60 period, though even then the magnitude of resource outflow from agriculture was still substantial.[3] In contrast to India, examined in the previous chapter, it appears that the agricultural sector in Taiwan made a net positive finance contribution to the rest of the economy throughout the period under study.

What particularly stands out in comparison to India is the relatively high share of producer goods in total purchases of the agricultural sector in Taiwan throughout the period. This share grew from about 23 per cent during 1911–15 to 39 per cent during 1926–30 and more than 43 per cent in the 1956–60 period. This expansion was due to the relatively slow growth of the consumption of the farm households, combined with faster increases in the inflow of new producer goods into the agricultural sector. Producer goods purchases increased rapidly from about 6 per cent of the agricultural value added during 1911–15 to about 19 per cent by the late 1920s. From then on, they grew more or less in line with the growth of the agricultural value added.[4]

The purchase of consumer goods, on the other hand, starting from the low rate of 20 per cent of the agricultural value added during 1911–15, increased gradually to reach the rate of 32 per cent by the end of the 1940s. In the post-war period, however, the rate declined once again to the very low levels of the early decades of the century. During 1950–5, for example, consumption goods purchases were about 23 per cent of the agricultural value added, increasing to only about 26 per cent by the early 1960s. These low ratios reflect the slow growth of consumption per head in the farm sector, which in real terms increased by an annual rate of 0.9 per cent during the entire 1911–60 period. The low rates of consumer goods purchases as a proportion of agricultural value added resulted to a large extent from the siphoning-off of a significant part of agricultural value added through

[2] The terms of trade effects are discussed in Section 6.4. As shown there, only in the post-war period do they constitute an important component of surplus outflow. However, even measured at the 1956–60 base year prices, the net resource flow estimates do not substantially differ from those reported in Table 6.1. As a ratio of total sales at 1956–60 base year prices, net surplus outflow was about 40% for the 1911–25 period, 18% for 1926–40, and 18% for the 1951–60 period.

[3] As can be seen from Table 6.1, in absolute terms, net resource outflow during this period was still higher than at any earlier time except 1950–5. During the 1956–60 period, net resource flow out of the agricultural sector was about 25% per cent of total agricultural sales.

[4] There were, of course, sharp fluctuations in the purchase/value added ratio, particularly during the Second World War period and its aftermath, but the overall long-term trend remained more or less stable after the 1930s.

taxes and rents by the absentee landlords. As we shall observe below, however, another important factor was the relatively high savings propensity of the farm households – particularly in the post-war period, when farm household incomes were supplemented by sizeable wage receipts from non-agricultural activities.

The level and the trends in agricultural sales, or the marketed surplus of the farm sector, depended on factors such as the institutional set-up of the farm sector and mechanisms of control and extraction of surplus product, the level and rate of growth of agricultural productivity, and the nature of technological progress in the sector. The ratio of marketed surplus to the value added in the agricultural sector increased from about 56 per cent during the 1911–15 period to an average rate of more than 71 per cent during the 1930s. Such high ratios reflected the relatively high productivity levels in Taiwan's agriculture, the effectiveness of surplus control mechanisms, and the high degree of commercialization of its agriculture during the first half of the century. During this period, agricultural taxation, and particularly rental payments to absentee landlords, acted as effective mechanisms of surplus transfer from the agricultural sector. In the post-war period and after the land reform programme, the income share of the farm households increased rapidly, and the marketed surplus ratio experienced a downward shift along with the growth in the consumption of the farm households. The marketed surplus ratio declined to about 60 per cent over the 1950–60 decade.

The level and the trends in different components of inflow and outflow of resources shown in Table 6.1, as well as the overall net resource flows, were determined by an interdependent set of variables on both the real and the financial sides of the economy. On the real side, high levels of agricultural productivity and the fast rates of technological progress in agriculture created the potential for a high and sustained rate of surplus outflow from agriculture. The financial mechanisms of resource flow, which changed over time in accordance with the changing institutions in agriculture and government policy, determined the actual magnitude and the form of surplus flow. In the following two sections of this chapter we shall discuss these determinants in turn.

6.3 Agricultural Growth and Surplus Transfer

The magnitude of the net resource contribution of agriculture to other sectors of the economy clearly depends on the rate of growth of its output over its inputs, particularly those purchased from the other sectors of the economy. As we have noted in Chapter 3, the higher the growth of productivity of labour in agriculture, the greater the possibility of resource outflow exceeding the inflow of consumer goods to the sector. Such productivity growth is normally associated with the increase in the inflow of new producer goods into the sector in the form of fixed and variable capital investment. The higher the increment in output resulting from the inflow of a composite unit of producer goods, the higher the potential for increasing net resource outflow from the sector.

The productivity of labour in Taiwanese agriculture, as shown in Table 6.2, grew by an average annual rate of 1.9 per cent during the 1911–60 period. What is significant is that these productivity growth rates were achieved despite a high rate of population growth and no possibility of major additions to land under cultivation. The rate of growth of population increased from about 1 per cent per annum during the first two decades of this century to 2.5 per cent per annum during the 1926–40 period and to more than 3 per cent per annum in the post-war period.[5] Because of the absorption of surplus labour in the non-agricultural sector, however, agricultural labour force had a much slower rate of growth.[6] Employment in the non-agricultural sector grew by more than 3 per cent per annum over the 1911–60 period as a whole, while agricultural labour force grew by 0.73 per cent a year. Nevertheless, given the limited possibilities of increasing land under cultivation, there was still growing population pressure on land. In the period before 1930, cultivated land per unit of labour was increasing somewhat, as population growth was still low and there were possibilities of extending the agricultural land frontier. During the 1930s cultivated land/labour ratio started to decline, but in the 1950s it stabilized at about 0.5 hectares. The rapid growth of agricultural labour productivity was sustained through the constant introduction

Table 6.2 Summary statistics for Taiwan's agriculture, 1911–1960

Year	Output per worker[a]	Per capita farm household income	Per capita consumption	Agricultural population	Agricultural labour force
1911	156	49	46	2106	1106
1915	148	46	44	2240	1165
1920	172	50	47	2279	1140
1925	238	72	63	2322	1152
1930	258	72	67	2512	1212
1935	289	82	75	2746	1325
1940	290	81	71	2955	1400
1950	278	87	75	3939	1731
1955	327	90	66	4546	1737
1960	385	95	72	5174	1754

[a] Value figures are in $Tm.
[b] Figures for the agricultural population and labour force are in thousands.
Source: Lee (1971, p. 13).

[5] In fact, it is remarkable that labour productivity seems to have accelerated with the increase in the population growth rate. The rate of growth of output per worker increased from 1.6% per annum during the 1911–40 period to more than 2.5% per annum during the high population growth period of 1950–60.

[6] As shown in Table 6.2, agricultural labour force increased from about 1.1m. in 1911 to 1.4m. in 1940. During the 1940s, as a result of considerable immigration into Taiwan, it jumped steeply by about 0.3m., and remained at 1.7m. for the rest of the period.

of technological innovations of a land-saving type, and through the efficient use of capital investment in the sector.

Technological change in agriculture during the first three decades of the present century was mainly the result of Japan's colonial policy of using Taiwan as its granary. The late 1910s witnessed a major land infrastructure and irrigation programme financed by the colonial government, which paved the way for the introduction of seed-fertilizer technology that made sustained increases in land and labour productivity possible. The introduction of new varieties of rice and sugar cane, destined for the Japanese market, had a major impact on agricultural productivity until the late 1920s. From then on, increasing use of modern chemical inputs and a constant diversification of agricultural production helped to maintain the momentum of agricultural productivity growth. In addition, investments in irrigation, flood control, and drainage allowed more intensive multiple cropping, thereby effectively increasing the land/labour ratio despite the shortage of cultivable land (Kikuchi and Hayami, 1985).

The significance of fixed investment in irrigation, land reclamation, and flood control in increasing agricultural productivity has been emphasized by various studies (see e.g. Rada and Lee, 1963; Lee, 1971, 1974; Hayami and Ruttan, 1971, pp. 205–10). An interesting aspect of Taiwan's experience, however, is that the share of fixed investment goods in the flow of producer goods into agriculture was very small until the late 1950s – it was less than 10 per cent of the value of producer goods and less than 3 per cent of the value of total goods purchased by agriculture. This reflected the significant use of internal resources of the farm sector for investment, especially surplus labour, and the extremely efficient use of capital investment in the agricultural sector. More than 90 per cent of the inflow of producer goods into agriculture was of the land-augmenting seed-fertilizer technology type, which, being perfectly divisible, was particularly suitable for the small operational farm units in Taiwan.

Another aspect of the growth of the agricultural sector, with important implications for the intersectoral resource processes, was the increase in the productivity of factors of production. These increases could result from such things as the reorganization of agricultural production, intensification of work processes, new and more productive combinations of existing inputs, or the application of genuine innovations based on new scientific discoveries. It is often difficult to separate the effect of each of these factors on total factor productivity. Given the heterogeneity of the factors of production, the reliance on total factor productivity as a single measure of technological progress is itself problematic and only possible under highly restrictive assumptions. The estimates produced by Lee (1971) suggest that about 60 per cent of output growth over the 1911–60 period was explained by technological change as measured by total factor productivity. Setting aside the problems associated with the restrictive assumptions used in deriving this estimate,[7] compared to similar estimates for other countries (e.g. India, discussed in the previous chapter), it does indeed show a high degree of agricultural productivity growth. This in turn explains the possibility of extracting the considerable surplus from agriculture, as noted in Section 6.2 above, as agricultural growth need not impose a heavy burden on other sectors in terms of resources.

This point can perhaps be more clearly demonstrated, without the need to resort to the strong assumptions involved in the measurement of total factor productivity estimates, if we consider the productivity growth rates for each production factor separately. To start with the producer goods purchases from the other sectors, as noted in Section 6.2 above, there was a rapid increase in the ratio of such inputs per unit of agricultural value added from about 6 per cent during the first decade of the century to about 19 per cent towards the end of the 1920s, when it stabilized at this level until the end of the study period. The acceleration in the producer goods purchase ratio during the first two decades of the study period coincided with the transition from traditional agriculture, based on extensive farming with little backward linkages with other sectors, to one based on intensive farming, where land yields constituted the major component of output growth (Mellor, 1973; Kikuchi and Hayami, 1985). The stability of the purchased input/value added ratio during the latter period indicated that the growth of output, and hence the potential contribution of agriculture to other sectors, was in line with the growth of its input requirements from the other sectors. In addition, as we have already noted, the new inputs led to high and, indeed, accelerating growth in both yields and the productivity of labour. In turn, labour productivity growth meant that consumption requirements, and hence purchase of consumer goods from other sectors per unit of agricultural output, could be kept at bay. Of course, the extent to which this could be done depended on the distribution of income and the control over the new income generated through productivity growth.

Despite the fast growth of labour productivity in agriculture, the per capita consumption of the agricultural population grew at a relatively slow pace, by 0.9 per cent per annum. This was partly due to the fact that the agricultural population grew faster than than labour, and partly to the siphoning-off of a large part of agricultural value added through rents by non-farming landlords, government taxation, and adverse terms of trade movements. It is to these issues that we turn now.

6.4 Financial Mechanisms of Surplus Transfer

Table 6.3 shows the financial channels of surplus transfer from the agricultural sector in two representative periods of the colonial and post-colonial era. As can be

[7] The assumptions are neutral technological change, constant returns to scale, and diminishing returns to production factors. In addition, it is assumed that factor remuneration is equal to the marginal product of the factor for all factors of production.

Table 6.3 Financing of net agricultural resource outflow, Taiwan, 1911–1960 ($Tm., current prices, annual averages)

	1931–5	1956–60
Net resource outflow (R)	63	948
(*a*) Agriculture's sales (X)	208	9665
(*b*) Agriculture's purchases (M)	146	8716
Financing items		
1. Net outflow of factor income ($F_a - Y_f$)	47	−813
(*a*) Land rents	56	739
(*b*) Labour incomes	−9	−1552
3. Net outflow of current transfers ($T_{fg} - T_{gf}$)	17	1446
4. Net outflow of capital transfers	−1	316
(*a*) Private ($K_{fo} - K_{of}$)	0	381
(*b*) Public ($K_{fg} - K_{gf}$)	−1	−65
5. Errors and omissions[b]	—	−1

[a] Refers to government's net transfers only.
[b] Mainly consists of private transfers ($T_{fo} - T_{of}$).
Source: Lee (1971).

seen, there have been significant changes in the structure of financial flows between the two periods. In the colonial period land rents constituted about 90 per cent of net resource outflow from agriculture. Up to the 1930s, these were mainly invested in financial assets in Japan. The financial surplus of the agricultural sector in this period showed in the large balance of payments surplus in Taiwan's foreign trade with Japan. In other words, Taiwan's agriculture was financing part of investment in Japan through its trade surpluses. During the 1930s, and particularly in the post-war period, this surplus was increasingly utilized in financing industrial investment in Taiwan itself. The inflow of labour income during the colonial period was thus relatively small, and the net flow of factor incomes was dominated by the outflow of rents. In addition to their sizeable contribution to net surplus outflow, rents also played an important part in terms of the marketed surplus of the farm sector. According to the estimates by Lee (1971, table 9), for example, the landlords' share of the marketed surplus of rice was about 57 per cent in the period 1911–5, increased to 66 per cent during 1916–20, and gradually declined to 42 per cent by 1936–40.

After rents, government taxes formed the second most important source of resource extraction from agriculture in the colonial period.[8] As Table 6.3 shows, net current official transfers formed about 27 per cent of the net outflow of resource from the farm sector during the 1931–5 period, and this figure is in fact

[8] Land taxes, constituting about 30% of government tax revenue, were the main item in agricultural taxation up to the 1940s. With the reform of the tax system in the 1940s, which put greater emphasis on income taxes, land taxes declined to about 7% of total tax revenue (Lee, 1971).

very close to the average for the whole of the 1911–40 period.[9] Despite the significance of government investment in agricultural development during this period the net consolidated government accounts (both current and capital) showed a sizeable and, indeed, growing net outflow of resources from agriculture. In other words, the outflow of resources through taxation was far greater than the inflow through capital investment and subsidies, and the difference was growing over time. This is in sharp contrast to the case of India, discussed in the previous chapter, where, due to the extremely low taxation of the agricultural sector, the flow of resources on government accounts was the reverse. What particularly stands out here is the higher tax burden on the farm households compared to the other sectors of the economy. As Table 6.4 shows, tax burden on agriculture in 1933 was heavier than on non-agriculture for all income groups. According to Lee (1971), tax burden was gradually shifting towards non-agriculture with the growing industrial development of the country, and therefore the relative tax burden on the farm sector during earlier years must have been heavier than that depicted in the table.

In the post-war period the sizeable financial outflows from the farm sector continued, but there was a significant change in the composition of the financial flows. Government taxes replaced rents as the main source of extraction of surplus from the agricultural sector (see Table 6.3 and Figure 6.1). Agricultural taxes in the post-war period mainly took the form of compulsory sales of rice and high fertilizer prices in compulsory barter exchange with the government.[10] The share of rents declined after land rents were reduced by legislation from an average of about 50 per cent to a maximum of 37 per cent, and with land reform, which

Table 6.4 Tax incidence of agriculture and non-agriculture by income group in Taiwan, 1933 ($T, at current value)

Income groups	Taxes paid by farmers	Taxes paid by non-farmers	Average
Below 400	49	47	48
400–800	113	87	100
800–1200	157	158	157
1200–2000	333	234	283
2000–3000	522	412	467
3000–5000	923	624	773
5000–7000	1417	833	1125
7000–1000	2018	1661	1840

Source: Lee (1971).

[9] For example, as a ratio of net surplus outflow, taxes were 26% during 1911–15, decreased to 20% during 1916–20, increased to 30% in 1921–5, decreased to 26% in 1926–30, and increased to 29% in 1936–40 (Lee, 1971, table 16).

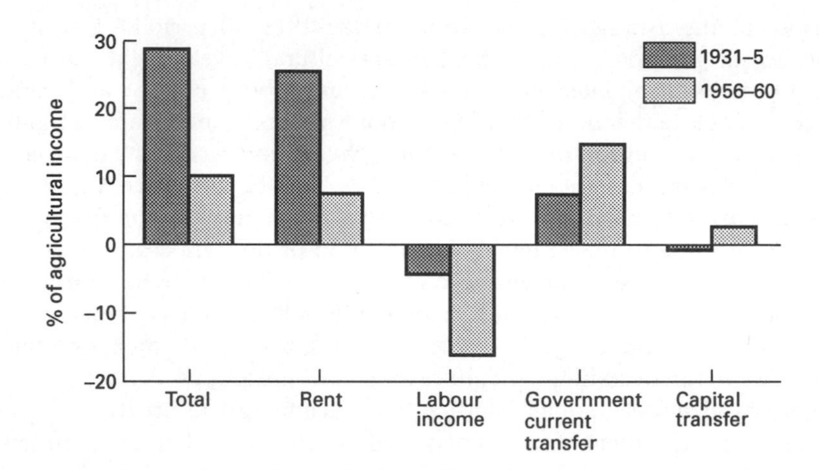

FIG. 6.1 Mechanisms of resource outflow from Taiwanese agriculture, 1931–5 and 1956–60

increased the proportion of land under owner cultivation.[11] The contribution of rents to resource outflow, which was three times that of government taxation during the colonial period, fell to 50 per cent of government taxation in the post-war period. Government taxation was instrumental in keeping up the pace of resource outflow from agriculture after the land reform of the early 1950s.

Another major change in the financial flows in the post-war period was the significant increase in the farm labour income from non-farming activities. This was more than twice the rent payments and easily offset the entire tax bill paid by the farm sector to the government (see Table 6.3). As a result, the balance of net factor income in the post-war period turned into a sizeable inflow. However, this was partly compensated for by net private capital outflows and, to a larger extent, by the substantial tax intake of the government from the agricultural sector. The considerable increase in private capital outflow from the farm sector through the credit system is another important aspect of the change in the financial flows in the post-war period, which is closely connected to the growth of incomes in the farm sector. As can be seen from Table 6.3, net private capital outflow increased from zero during the colonial period to T$ 381m. during the 1956–60 period, equivalent to more than 40 per cent of the total net financial outflow from agriculture. What this signifies is that, with the growth of income in the farm sector arising from fast productivity growth in agriculture and the increase in employment opportunities

[10] During the 1950s, the government collected more than 50% of the rice sold and more than 30% of rice production. See Lee (1971, pp. 80–5).

[11] With the 1953 land reform programme, about 60% of private tenanted land was purchased by the government and resold to 200,000 tenant families, who then became independent owner operators (Lee, 1974).

in the non-farm sector, surplus outflow could take place automatically as a result of the increase in the savings of the farm households and through the credit institutions. Of course, the magnitude of the outflow depends on the savings propensity of the farm households and the degree of development of the financial institutions in the countryside.

The data provided in Table 6.3 also make it possible to measure the 'savings surplus' of the farm sector. In Chapter 2 (Equations 2.10 and 2.11) the savings surplus of the farm sector was defined as the net financial surplus minus net factor incomes and current transfers, which equals the net capital transfer of the farm sector. As can be seen from Table 6.3 (row 4), the savings surplus of the farm sector during the colonial period turns out to be negative, and in the post-war period, though positive, it is no more than about 30 per cent of the total net finance contribution of agriculture. The example of Taiwan brings out the criticisms put forward in Chapter 2 against regarding the savings surplus as the contribution of agriculture to capital formation in other sectors of the economy. Rents and government taxes on agriculture should clearly be counted as an outflow from agriculture, and hence a contribution of agriculture to capital formation in other sectors. Otherwise, focusing on the savings surplus alone, one may get the mistaken impression that an agriculture heavily squeezed through taxation and rents makes little or, in fact, a negative contribution to other sectors – as in the case of Taiwan during the colonial period.

The Terms of Trade Effect

The channels of financial resource transfers discussed above were referred to in Chapter 2 as the visible accounts. A further mechanism of resource transfer was through the terms of trade changes, i.e. the invisible account. During the colonial period, agricultural terms of trade fluctuated, mainly as a result of the Japanese government's policies regarding imports and pricing of rice in the domestic Japanese market. The overall trend, however, was an upward movement in favour of agriculture until the end of the 1920s (see Table 6.5). During the 1930s the terms of trade started out by moving against agriculture, but recovered somewhat towards the end of the decade. In the post-war period there was a sharp drop in terms of trade against agriculture, which, as we have already observed, was due to the government's pricing policies in its compulsory rice collections and fertilizer sales. This amounted to an important source of surplus outflow from agriculture in the 1950s.

A decomposition of the real net resource outflow from agriculture into its visible and invisible components is given in Table 6.6. The invisible component refers to the income terms of trade gains in non-agriculture, and the visible component is the real value of the financial flows discussed above, in 1935–7 prices. The table shows a clear trend towards improving income terms of trade for agriculture up to

Table 6.5 Terms of trade of agriculture, Taiwan, 1911–1960

Year	Price indices		Net barter terms of trade (P_a/P_n)
	Agricultural sales (P_a)	Agricultural Purchases (P_n)	
1911–15	60	73	82.19
1916–20	92	119	77.31
1921–5	102	114	89.47
1926–30	103	103	100.00
1931–5	80	86	93.02
1936–40	120	123	97.56
1951–5	141	177	79.66
1956–60	248	298	83.22

Note: 1935–7 = 100.
Source: Lee (1971).

the 1930s, and a sharp reversal thereafter. The income gains due to the terms of trade improvements during the first three decades, however, were more than compensated for by the increase in the real value of the outflows on the visible account – that is, primarily the increase in the real value of the rent and tax receipts. For example, the T$ 16.3bn. decline in the invisible outflows between the 1911–15 and 1926–30 periods was more than matched by an increase of about T$ 26m. on the visible account during the same period (see Table 6.6). This contrasts with the experience of India, particularly in the aftermath of the new agricultural strategy, where agricultural income gains remained largely untaxed.

This trend was sharply reversed from the mid-1930s, and the income losses to the agricultural sector due to the adverse terms of trade movements were particularly significant during the 1950s. During the 1956–60 period, for example, the income losses arising from the adverse terms of trade movement constituted more than 66 per cent of the total real net resource outflow from agriculture.[12] To get some idea of the relative orders of magnitude involved, it would be instructive to compare the income losses due to the terms of trade movement with the increases in the value added of the agricultural sector in the post-war period. In the period between the latter half of the 1940s and the first half of the 1950s, for example, agricultural value added increased by about T$ 90m., while the income losses from the adverse terms of trade movement were about T$ 53m., both valued at 1935–7 prices. In other words, the income squeeze on the farm sector through the terms of trade movement in the post-war period constituted more than 60 per cent of the increase

[12] Note that, despite the improvement in the barter terms of trade of agriculture during 1956–60 compared to 1950–6, the invisible outflows continued to increase as a result of the fast growth of purchases of the agricultural sector from the non-agricultural sector.

Table 6.6 Terms of trade and capital flows between agriculture and non-agriculture in Taiwan, 1911–1960 (1935–7 prices)

Year	Net barter terms of trade	Net real capital outflow[a]		
		Total	Visible outflow[b]	Invisible outflow[c]
1911–15	82.2	49.2	32.9	16.3
1916–20	77.3	71.2	43.0	28.2
1921–5	89.5	59.9	43.9	16.0
1926–30	100.0	59.2	59.2	0.0
1931–5	93.0	90.2	72.1	18.1
1936–40	97.6	89.5	82.1	7.4
1950–5	79.7	112.8	51.9	60.9
1956–60	83.2	96.1	31.9	64.2

[a] $Tm.
[b] Calculated as $(X - M)/P_x$.
[c] Calculated as $m_a(1 - P_m/P_x)$.
Source: Tables 6.1 and 6.5.

in incomes resulting from value added growth in the sector.[13] The post-war experience of Taiwan in this regard also stands in contrast to that of India, discussed in the previous chapter, where the terms of trade movements were shown to entail large income gains in the agricultural sector.

6.5 Concluding Remarks

The experience of Taiwan in terms of the intersectoral resource flows between agriculture and non-agriculture stands out as one of sustained and substantial surplus flow out of the agricultural sector throughout the period under study. A fast rate of productivity growth in agriculture was a necessary precondition for resource outflows on such a large scale. Effective government action both in transforming the technological basis of production and in siphoning off a large share of the fruits of productivity growth through various financial channels was instrumental in the process of surplus transfer from agriculture.

Technological transformation of the agricultural sector in Taiwan prior to the Second World War was strongly shaped by the needs of the Japanese colonial administration in procuring a sizeable marketed surplus for export to Japan. This made a crucial difference to the Japanese colonial policy in Taiwan, in contrast to other colonial rules at the time (e.g. the British rule in India), where the main focus of policy was on revenue administration and the extraction of financial surplus from agriculture rather than on the agricultural marketed surplus as such. It was

[13] This was only partially compensated for by a decline in the real value of the visible component. As can be seen from Table 6.6, the real visible resource outflow component declined by about T$ 30m. during the same period.

this urge to expand the agricultural marketed surplus and agricultural supplies in general that brought about the far-reaching transformation of the agricultural sector in Taiwan through direct intervention by the Japanese administration. According to Lee (1971), force often had to be used to introduce the new technological changes.[14] During the first two decades of this century, agricultural policy was mainly directed towards institutional reform, agricultural education, and moderate investment in basic infrastructure. During the 1920s, the growing population pressure on land and the increasing demand for rice and sugar in the Japanese market necessitated big increases in land productivity and the transformation of the technological basis of Taiwanese agriculture. The 1920s witnessed heavy investment by the government in irrigation and land improvement, and the move towards intensive farming based on increasing use of fertilizers and new varieties of seeds. By the mid-1920s, about 60 per cent of the land under cultivation was already irrigated. This, together with other infrastructural investments during the early decades of the century, laid the foundations for the rapid spread of the new crop technologies and productivity growth in agriculture in the subsequent period.

Surplus outflow from agriculture after the mid-1920s was particularly facilitated by the fact that the process of technological change and productivity growth in agriculture did not divert a disproportionately high amount of resources from the other sectors of the economy. Fixed investment in irrigation and land improvement drew largely on the existing surplus labour within agriculture, and most of the producer goods inputs purchased from the other sectors were of the working-capital type, mainly composed of inputs such as improved seeds, chemical fertilizer, and pesticides. Technological change was based on improved labour-using practices, and more than 90 per cent of the inflow of producer goods into agriculture was of the land-augmenting seed-fertilizer type. The rate of growth of the inflow of producer goods purchases into the agricultural sector was particularly high during this period. However, as a result of the high efficiency of resource use and the rapid rate of technological change, the output response was commensurate with the rate of increase in inputs. It was this high rate of return on the use of new inputs which created the potential for the sustained extraction of a sizeable surplus from agriculture for such a long period of time.

The high rates of surplus extraction through government taxation, rents, and adverse terms of trade movements kept the rate of consumption in the farm sector low, and helped to realize the potential for resource outflow created by the rates of productivity growth. There was a radical shift in the financial mechanisms of resource transfer between the colonial and the post-colonial periods. During the colonial period the main channels of surplus outflow were through rents and government taxation. Terms of trade movements were mostly in favour of agricul-

[14] 'The process of altering old cultivation methods and the extension of use of new varieties [of rice] ... was not characterized by persuasion, but rather by government enforcement. Police stayed in the local communities and effectively participated in agricultural extension services.' (Lee, 1971, p. 41)

ture. In the post-war period and after the land reform of the early 1950s, rents lost their significance as a major mechanism of resource transfer, and government taxation and adverse terms of trade movements became the predominant sources of surplus transfer. During this period, also, the rapid rate of industrial growth and the opening-up of work opportunities outside agriculture brought about a significant inflow of labour income into the farm sector. The net balance during the 1950s, however, still showed an outflow from agriculture which in absolute terms was actually higher than at any other time during the colonial period. What stands out here is the ability to sustain such high rates of surplus flow during a period of rapid population expansion, with population growth rates of more than 3 per cent per annum during the 1950s. This is partly explained by high rates of employment generation in the non-agricultural activities, which reduced the rate of increase of population pressure on land, but the larger part of the explanation lies in the fast rates of land-augmenting technological progress, which sustained the rate of labour productivity growth in agriculture despite the high population growth rates.

The growth of farm household incomes in the post-war period also led to a relatively sizeable surplus outflow in the form of voluntary savings by the farm households through the credit system. This phenomenon, which, as we shall see in Chapter 8, was also important in the case of Japan, demonstrates that, in the context of a technologically dynamic agricultural system, surplus outflow need not necessarily be enforced through taxation or rent extraction. Of course, the magnitude of the outflow through this channel depends on the farm households' propensity to save and the degree of development of financial institutions in the countryside.

A final point to be emphasized in relation to the experience of Taiwan is that the substantial squeeze on agriculture through taxation, rents, and other mechanisms does not seem to have retarded the rate of technological advance and output and productivity growth in the sector. During the post-war period in particular, substantial adverse movements in the terms of trade were combined with the highest rates of growth in output and yields in the agricultural sector. This signifies the fact that, in a technologically dynamic agriculture, productivity growth can help to maintain relative profitability, and hence the inducement to invest in agriculture, despite the adverse terms of trade movements.

7

Intersectoral Resource Flows in Iran, 1963–1977

7.1 Introduction

In this chapter we examine the pattern of intersectoral resource flows in Iran during the 1963–77 period. Estimates of net product or net finance contribution of agriculture on the real side are not available for Iran. The predominant assumption in the literature, however, has been that, given the industrialization bias of the government and the relative neglect of agriculture, the direction of net resource flow must have been away from agriculture during the 1960–77 period (see e.g. Ashraf and Banuazizi, 1980; Katouzian, 1978). In an earlier work I estimated partial measures of resource flow, such as the marketed surplus and the net agricultural surplus from the real side, and income terms of trade and capital inflows from the financial side (see Karshenas, 1990*b*). In this chapter we utilize these partial measures in combination with new calculations of financial outflow in order to provide estimates of the direction of net resource flows over the 1963–77 period. The results confirm the conclusions of my earlier study, i.e. they indicate a net resource inflow into agriculture of a relatively large magnitude. Since this is in conflict with the conventional views in the literature on the direction of agricultural surplus flow, care has been taken to use the most conservative estimates of resource inflow when faced with possible errors due to inaccuracies of data. In this sense, the estimates mainly indicate the direction of net resource flow, with a bias towards underestimating resource inflow to agriculture.

The chapter is organized as follows. In the next section we discuss the new estimates of financial flows and compare the results with earlier estimates of agricultural surplus flows. This constitutes the visible component of the net financial resource flows. Section 7.3 discusses the terms of trade effect, or the invisible component of resource transfer. In Section 7.4 we consider the real side of the resource flow process, in order to identify the main underlying factors which explain the pattern of net intersectoral resource flows. A summary of the chapter and its conclusions are contained in Section 7.5.

7.2 The Direction of Net Resource Flows

Since accurate data for estimating the magnitude of resource flow from the real side are not available for all the years in the 1963–77 period, calculation here is attempted from the financial side. Table 7.1 shows different sources of inflow and outflow of funds in Iranian agriculture for the 1963–77 period. Row 1 of the table shows the inflow of funds through government development expenditure. Since there were no land taxes at this time and income taxes on farm households were virtually non-existent, this could be taken as the net resource inflow through official current and capital transfer accounts.[1] Of course, this is a conservative estimate, as neither the inflow of government current administrative expenditure in agriculture nor interest subsidies on cheap official loans have been included. The former has been omitted because of the objections which may be raised as to its inclusion, and the latter because of estimation problems and the ambiguities in determining the 'correct' or equilibrium market interest rate in Iran over this period.[2] Since we are mainly interested here in the direction of resource flows rather than exact estimates of its magnitude, and as the primary aim is to test the agricultural squeeze hypothesis, we preferred to err on the conservative side in estimating resource inflows into the agricultural sector.

The second row of the table shows the net private capital flows through the financial institutions. Since accurate data on net capital flows through the informal financial sector are not available, we have assumed a net annual balance on that account. This seems a plausible assumption, as informal sector credits are expected to be very short term and, if anything, any error involved would be biased towards an underestimation of the inflow of net credit to the farm sector. As can be seen, the inflow of credit through the banking system was increasing over time, and expanded particularly during the oil boom years of 1973–77. Though resource outflow through increase in cash in circulation and bank deposits was also growing rapidly in this period, the overall outcome, with the exception of the fourth plan period, seems to have been a net inflow on the private capital account.[3] As row 3 of the table shows, the combined official and private capital flows show a large and increasing inflow towards agriculture throughout the study period.

Estimates of factor income flows also indicate an increasing inflow towards the agricultural sector (row 4). These essentially consist of wage labour income inflow, interest on loans, and rent payments. The outflow of interest on loans has been calculated separately for bank loans and informal market loans. The agricultural

[1] Other indirect taxes and subsidies are, of course, taken into account in measuring the resource flow through terms of trade mechanism.

[2] See Salehi-Isfahani (1989) for a study of interest subsidies for the economy as a whole.

[3] The increase in cash in circulation in the farm sector is calculated by multiplying the increase in total cash in circulation outside banks by the ratio of agricultural marketed surplus to national income. A similar procedure was applied to the measurement of farm sector deposits in banks. This procedure tends to overestimate the farm household deposits, as the marginal propensity to save in the higher income non-farm sector is expected to be greater.

Table 7.1 Financing of net agricultural resource outflow, Iran, 1963–1977 (bn. rials)

	1963–7	1968–72	1973–7
1. Official capital[a] ($K_{fg}-K_{gf}$)	−36.6	−73.6	−244.0
2. Net private capital ($K_{fo}-K_{of}$)	−1.0	6.7	−16.7
Change in bank credit[b]	−8.9	−13.9	−148.9
Change in cash in circulation	2.1	5.0	24.6
Outflow via the banking system[c]	5.8	15.6	107.6
3. Net capital outflow (col. 1 + col. 2)	−37.6	−66.9	−260.7
4. Net factor income (F_a-Y_a)	−21.5	−44.9	−119.4
Interest on bank loans[d]	3.2	7.5	48.6
Interest on other loans[e]	15.4	29.9	67.8
Other income[f]	−40.1	−82.3	−235.8
5. Net resource outflow	−61.1	−111.8	−380.1
Visible flows in real terms[g]	−60.7	−104.0	−252.4
6. Net agricultural surplus[h]	97.8	271.4	606.7
7. Total gross capital inflow[i]	121.7	277.1	916.4
8. (Row 6−Row 7)	−23.9	−5.7	−309.7

[a] Development expenditure of the government.
[b] End-of-year change in credit granted by commercial and specialized banks to the agricultural sector.
[c] End-of-year changes in savings and sight deposits by agricultural households.
[d] Net interest payment to the banking system.
[e] Interest payment on informal sector loans.
[f] Inflow of income to the farm sector from non-agricultural activities, including interest on bank deposits, excludes rent payments.
[g] Row 5 deflated by agricultural purchase price index (1963 = 100).
[h] Agricultural value added minus consumption expenditure of direct producers.
[i] Row 1 plus gross inflow of loans to agriculture.
Sources: Karshenas (1990*b*), CSO (1976), and Bank Markazi Iran (1962–77).

sector at this time received an increasing amount of credit at highly subsidized rates of interest from government-financed agricultural credit institutions.[4] The interest payments on loans from the informal credit market are rough estimates based on the available information, which also ensures that the possible errors involved tend towards overestimation of the outflow.[5] The last item of the factor income accounts in the table refers to total gross inflow of income into the farm sector, largely formed by wage labour income, but also including interest and other factor income receipts.[6] As can be seen, these form an important source of resource

[4] For a description of the role of agricultural credit institutions over the study period, see Karshenas (1990*b*, ch. 4). Interest on bank credits in Table 7.1 has been calculated by applying interest rates on bank loans of average duration to the outstanding credit at the end of the year over the study period. The following interest rates were adopted: 6% for 1963–7, 7–8% for 1968–72, and 9–10% for 1973–7. The errors involved in this procedure are likely to lead to an overestimation of the outflow of funds from agriculture.

inflow into the farm sector, particularly during the boom years of the 1970s, when the phenomenal rates of growth of the non-agricultural sector, fuelled by oil export revenues, led to a rapid rate of increase of demand for labour.[7] The only major missing item from the factor income flow accounts is the outflow of income through rent payments to absentee landlords. However, with the land reform programme, which substantially reduced large-scale absentee landlordism, one would expect this to have diminished considerably as a source of outflow of funds from the sector.[8] In any event, this omission is unlikely to be significant enough to reverse the relatively large financial flows to agriculture through other mechanisms shown in the table.[9]

The financial flows, therefore, appear to indicate a net resource transfer into the agricultural sector of a rapidly increasing magnitude over the study period. This confirms the results of my earlier work, where a combination of real- and financial-side data was utilized to provide some indication of the direction of resource flows (see Karshenas, 1990*a*; 1990*b*). This was done by a comparison of net agricultural surplus estimated from the real side, and the gross inflow of funds into agriculture

[5] The main missing information is the stock of outstanding credits from the informal sector. This was estimated by assuming the stock of credit from the informal sector to bear the same ratio to credits from the banks as the ratio of gross loans from the informal sector to those from the banks. The latter ratio was derived from Karshenas (1990*b*, table 6.5). This procedure is expected to lead to an overestimation of informal credits, as these are very short-term loans and are expected to be repaid within a year. An interest rate of 30% was assumed throughout the period for such loans. This assumption is also on the high side, as interest rates in Tehran bazaar were, for example, only 15.5% in 1974, increasing to 21% in 1976 and a maximum of 25% in 1977 (Salehi-Isfahani, 1989).

[6] This was estimated on the basis of household budget surveys. According to the household budget surveys for 1975 and 1976, non-agricultural income as a share of the total income of agricultural households remained stable at around 12.6% for these 2 years. One would obviously expect this share to have increased noticeably during the oil boom years of the 1970s. We thus assumed a 5% share in 1963, increasing linearly to 12.5% in 1977. These estimates are clearly on the conservative side.

[7] The 1970s were a period of relatively high rates of growth of real incomes in rural areas, and particularly amongst the poorer agricultural classes. The period witnessed a strong boom in side activities, such as carpet-weaving and brick-making, in rural areas (Majd, 1983), as well as in urban construction activity, which drew heavily on rural migrant labour. Over the 1970–7 period, real wages of unskilled construction workers rose by an annual rate of 19.5%.

[8] Of course, rent payments were replaced by instalments which the peasant farmers who received land under the land reform programme had to pay to the government as the price of their newly acquired holding. The amount paid by land recipients was negligible, however, compared to other sources of flow of funds shown in Table 7.1. For example, during the 1962–70 period, the government paid about 7.3bn. rials to the landlords in partial payment of acquired land, but the amount paid by the recipient farmers to the government during the same period was only 0.49bn. rials. See Denman (1973, ch. 12) for further details.

[9] For example, to reverse the total net inflow of resources shown in Table 7.1 (row 5), rent payments to absentee landlords should be at least 62% of net agricultural surplus in 1963–7, and 63% in 1973–7. Net agricultural surplus is defined as the value added in agriculture minus the consumption of direct producers (row 6). Clearly, these ratios are too high, particularly for the 1970s, by which time the land reform programme was completed.

from the financial side. We have reproduced these variable in rows 6 and 7 of Table 7.1. If we recall the discussion in Chapter 2, the net agricultural surplus was defined as:

$$NS_a = F_a - (C_{af} + C_{nf}) = (X_a - M_a) + (I_{af} + I_{nf}),$$

or

$$NS_a = F_a - C_a = (I_a + X_a - M_a),$$

where the variables are defined as in Chapter 2. Net agricultural surplus refers to resources made available by the agricultural sector for investment within the sector itself and utilization by other sectors. The difference between this notion of agricultural surplus and that of the net finance or net product contribution of agriculture is thus the total investment in agriculture. From this we can get the following measure of the net product contribution of agriculture:

$$R = X_a - M_a = NS_a - I_a,$$

that is, the net resource outflow from agriculture is equal to net agricultural surplus minus total investment in the agricultural sector. Since data on total investment in agriculture are not available, the earlier work tried to find a proxy for it from the financial side. As far as the government investment in agriculture is concerned, official capital inflow, that is, the government development expenditure in agriculture, could act as a reasonable proxy. For private sector investment, we assumed total gross loans to agriculture to be equal to private investment, which is clearly a strong assumption. With these assumptions, the net surplus transfer to agriculture would be equal to the net agricultural surplus minus total gross capital inflows to agriculture, which is shown in row 8 of Table 7.1. As a comparison of these figures and the new estimates in row 5 shows, this measure, though producing the same results on the direction of resource flows, clearly underestimates the magnitude of the surplus inflow, particularly during the 1960s. This indicates that, barring other errors and omissions, private investment in agriculture during the 1960s was well above loan capital inflows, but during the 1970s, with the rapid increase in bank lendings, the two measures come reasonably close.[10]

So far we have been discussing the resource flows on the visible component of the financial flow accounts at current prices. As Table 7.1 shows, these results do not change substantially when converted into real values. Even in real terms, i.e. at 1963 base year prices, the visible account shows a growing resource inflow into the agricultural sector during the study period. These figures, however, do not take into account the invisible flows on the financial side, that is, the income flows through the terms of trade effect.

[10] Another source of underestimation in the latter approach may be an upward bias in the estimates of net agricultural surplus. As pointed out in Karshenas (1990*b*, appendix A1), there is a possible upward bias in these estimates due to an underestimation of rural household consumption in Iranian national accounts for this period.

7.3 The Terms of Trade Effect

Estimates of net barter and income terms of trade of the agricultural sector for the 1963–77 period are shown in Table 7.2. As can be seen, there was a sustained terms of trade movement in favour of agriculture in this period, which accelerated during the oil boom years of the 1970s. The net barter terms of trade improved in favour of agriculture by more than 57 per cent during the period as a whole, more than 47 per cent of which took place in the 1970–7 period. The income gains due to the terms of trade improvements in the agricultural sector by 1977 were equal to about 82bn. rials (in 1963 prices), which was more than the real income gains arising from the normal growth of value added in the sector. Added to resource inflows through the budgetary, credit, and other mechanisms on the visible account, this does indeed imply a substantial net resource inflow into agriculture. This result, however, crucially depends on the choice of the base year for the measurement of the terms of trade effect. Given the substantial improvement in the agricultural terms of trade, taking one of the later years as the base year can easily reverse the above conclusion.

In choosing an appropriate base year, it should be noted that during the study period, the Iranian government followed a strong import substitution policy which afforded a high degree of protection to the industrial sector. Similarly, the prices of major agricultural products were regulated by the government through its food import policy. In addition, various consumer goods and producer goods inputs into agriculture received large price subsidies from the government throughout the

Table 7.2 Terms of trade of agriculture, Iran, 1963–1977

Year	Price indices		Net barter terms of trade (P_a/P_n)	Income terms of trade	
	Agricultural, sales (P_a)	Agricultural purchases (P_n)		Base year 1963	Base year, 1970
1963	100.0	100.0	100.0	0.0	−6.9
1964	110.4	102.0	108.2	5.2	−1.4
1965	110.6	100.5	110.0	6.7	−0.2
1966	109.5	100.0	109.5	7.1	−0.6
1967	109.1	100.9	108.1	6.5	−1.9
1968	110.6	102.6	107.8	7.1	−2.4
1969	114.6	104.3	109.9	9.6	−0.4
1970	116.7	105.8	110.3	10.7	0.0
1971	131.3	110.5	118.8	17.4	8.4
1972	137.9	114.4	120.5	21.6	11.4
1973	159.0	126.4	125.8	30.3	19.3
1974	211.4	139.7	151.3	57.5	48.6
1975	205.9	145.3	141.7	53.3	42.5
1976	237.7	161.2	147.5	65.7	54.4
1977	283.7	180.4	157.3	81.7	70.9

[a] Measured as x_a ($P_x/P_m - 1$)bn. rials, at 1963 and 1970 prices.

Source: Karshenas (1990*b*).

period. Unfortunately, there are no empirical studies which could show the degree of protection of the agricultural sector during different years of the study period. Judging by Ishikawa's notion of a 'normal' year (see Chapter 2), the post-oil boom years of 1973–7 are obviously not representative years, as the economy was in severe disequilibrium as a result of the expenditure of the huge oil revenues in the domestic economy. Taking a year in the early 1970s as the base year, however, does not dramatically change the results. For example, the income terms of trade measured at 1970 base year show a relatively small outflow during the 1960s – negligible relative to the inflow of financial resources on the visible account – turning to an inflow during the 1970s of a similar magnitude to income gains measured at 1963 prices (see Table 7.2).

A comparison between the movement of the agricultural terms of trade in Iran and price movements at the international level can be also useful in judging an appropriate base point price system. Figure 7.1 shows the movement of these two terms of trade series over the 1963–77 period (both depicted as index numbers with 1963 = 100).[11] As can be seen, with the exception of the world commodity price boom years of 1973 and 1974, the agricultural terms of trade in Iran exceeded the world level, with a widening gap over time. By the end of the study period the terms of trade in Iran were more than 65 per cent higher than the international terms of trade. Of course, the degree to which this indicated a relative overpricing or underpricing of agricultural as opposed to industrial products in Iran depends on the degree of protection of various sectors and the price subsidies given to agriculture in 1963, on which adequate information is not available. The available information, however, suggests that the relative underpricing of agricultural products in 1963 is unlikely to have been greater than the more than 60 per cent wedge between the world and Iranian terms of trade in the early and the late 1970s. For example, according to the data provided in Karshenas (1990*b*, ch. 5), the rate of nominal protection of industrial products, as measured by the ratio of customs proceeds to total imports in the early 1960s was less than 30 per cent. Considering that the highest rates fell on what were considered to be luxury consumer products which did not figure at all in the purchases of the agricultural sector, the impact on agricultural terms of trade would have been even less pronounced.[12]

[11] The world agricultural terms of trade series here, as in other country study chapters, is based on the data in Grilli and Yang (1988). It is an index of agricultural commodity exports divided by exports of manufactures by industrialized countries, both measured on fob basis. In constructing the agricultural commodity price index, a weighting system of 2 : 1 was adopted for food and non-food commodity price series respectively, similar to the weights of these two commodity categories in total sales of the agricultural sector in Iran.

[12] Of course, the degree of protection of the industrial sector in itself does not imply relative underpricing of agricultural products, as agricultural imports were also under strict government control. It is often argued in the literature that government subsidies on the sale price of its grain imports implied negative protection for the agricultural sector. This need not be true, as imports were a small part of total supplies (about 3%) during the 1960s. In addition, it should be pointed out that agriculture received many of its inputs at subsidized prices. The net result of these policies on

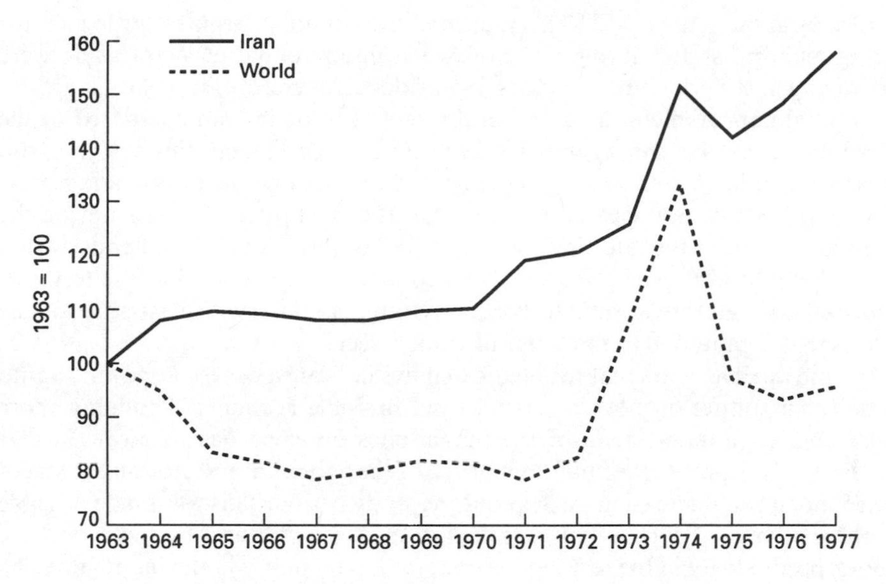

FIG. 7.1 Agricultural terms of trade, Iran and the world, 1963–1977

To examine the degree to which the agricultural terms of trade in Iran in 1963 were depressed compared to a free trade situation, it would also be instructive to consider the historical experience during the 1950s. From the Korean War commodity price boom of the early 1950s onwards, the agricultural terms of trade at world level followed a downward path. According to the data provided in Grilli and Yang (1988), the international terms of trade of agricultural primary commodities relative to industrial products declined by about 10 per cent between 1953 and 1960. Though estimates of agricultural terms of trade for this period in Iran are not available, the existing evidence suggests that terms of trade moved substantially in favour of agriculture and that the sector benefited from a higher degree of protection than industry (Karshenas, 1990*b*, ch. 5). For example, the wholesale price indices of food grain and animal foods increased by 67.1 and 103 per cent respectively between 1953 and 1960, while the general index of wholesale prices increased by only 15 per cent over the same period. This was in fact indicative of a decline in the price of manufactured goods over the 1953–60 period. The wholesale price index of imported goods (largely consisting of industrial products) declined by 17 per cent over this period, and the wholesale price index of

agricultural terms of trade in 1963 is not clear, but, as the relative movement of terms of trade in Iran and the world shows, the degree of protection afforded to the agricultural sector was growing during the 1963–77 period, contrary to what is commonly believed.

textiles by about 5 per cent.[13] This indicated a substantial terms of trade gain for the agricultural sector during a period when international terms of trade were moving against agriculture. The existing evidence, therefore, seems to suggest a substantial improvement in agricultural terms of trade in Iran compared to the international market throughout the 1953–77 period. Though this evidence still does not establish which year comes closest to approximating a free trade regime relative price system, it clearly indicates that the food price increases during the economic boom of the late 1950s would have been much more moderate with a more liberal food import policy. Calculated at 1963 prices, the income terms of trade would therefore appear to be, if anything, a conservative estimate of the 'true' income gains due to the terms of trade effect.

To sum up: it appears that the financial flows indicate a net resource inflow into Iranian agriculture on both the visible and invisible accounts. Though a more exact and accurate estimate of the magnitudes involved has to await further research, it is clear that the financial burden of the agricultural sector on the rest of the economy was increasing throughout the study period. This was made possible by the existence of substantial surpluses in the oil sector and by the growth of labour productivity in the rest of the economy, particularly in the manufacturing sector. To investigate the underlying factors which could help explain this phenomenon, we need to consider the real side of the resource flow process, which is the task of the next section.

7.4 Resource Flows on the Real Side

The financial flows discussed in the previous section have their counterpart in commodity flow accounts on the real side. The substantial inflow of funds through the development budget, private capital accounts, wage labour income, and the income gains through the terms of trade effect were indeed translated into a large gross inflow of resources on the real side into the agricultural sector over the 1963–77 period. The flow of modern technical inputs such as chemical fertilizers, pesticides, and agricultural machinery and implements into the agricultural sector grew rapidly over this period. The use of chemical fertilizers expanded from about 32,000 tons per annum in the early 1960s to 675,000 tons by the mid-1970s, while investment in agricultural machinery experienced a sevenfold increase in real terms over the same period. In order to understand why such substantial gross financial and real inflows did not over time create a commensurate rate of financial outflow,

[13] The reason for the decline in manufacturing prices in this period was the removal of trade restrictions and the reduction of tariffs following the resumption of oil exports in 1954, which led to a rapid increase in the imports of manufactured goods. Food imports during this boom period were kept almost constant, and, as a result, agricultural product prices increased substantially. For more details, see Karshenas (1990*b*).

we must examine the efficiency of resource use on the real side, in particular with respect to productivity of investment in agriculture and consumption of the farm sector.

A comparison of the similarities and differences between the Iranian experience and that of Taiwan (see Chapter 6) in this regard is instructive. Like Taiwan, the Iranian economy witnessed rapid population growth, close to 3 per cent per annum, over the 1963–77 period. Similarly, the number of agricultural labourers in Iran rose relatively slowly over this period, at less than 1 per cent per annum, as a result of the high rates of rural–urban migration and the rapid absorption of labour in other sectors of the economy.[14] Agricultural value added grew by moderate rates of nearly 4 per cent per annum, and a labour productivity growth rate of 2–3 per cent – in the same range as Taiwan – was achieved. The similarities between the two economies, however, stop here. The contrasting experiences in the direction of net resource flows in the two economies resulted from the differences in the efficiency of capital usage as well as in consumption behaviour in their respective agricultural sectors.

The main thrust of agricultural policy in Iran over the study period was towards highly capital-intensive mechanized farming. Apart from giving encouragement to the private sector for land consolidation and mechanization through various forms of subsidies (see Nowshirwani, 1976), the government also intervened directly to hasten the mechanization process by creating new forms of production units such as farm corporations and agro-business enterprises. Government development expenditure by and large benefited only the new large-scale mechanized farms, which made little contribution to the growth of output and were created at the expense of the displacement of thousands of peasant households.[15] More than 30 per cent of the credit granted by the specialized banks was also absorbed by the large-scale state-sponsored agro-business concerns. A large body of research on Iranian agriculture testifies to the lack of commensurate response of agricultural production – and particularly production in the new government-sponsored

[14] The high rates of rural–urban migration in this period are reflected in the differential rates of growth of rural and urban populations. While urban population grew at an average annual rate of 4.9% during the 1966–76 period, the rate of growth of rural population over the same period was 1.1% per annum (Karshenas, 1990*b*, app. P1).

[15] See e.g. ILO (1972), Katouzian (1978), Shafa-eddin (1980), Ashraf (1982). According to the ILO report, 'large private capital (e.g., agro-industrial companies) is given the most assured supply of irrigation water, lands and government protection and subsidies (specially investment in infrastructure), most of which is denied to the smaller private individual entrepreneurs in agriculture.' For example, the bulk of development expenditure allocated to agriculture during the 1960s (twice the expenditure on other sectors of agriculture) was utilized for dam construction, and the lands made available below the dams were handed over to mechanized agro-businesses. During the 1970s, also, more than 40% of development expenditure in agriculture (excluding irrigation) was directly allocated to agro-businesses and farm corporations.

projects – to the considerable human and financial resources allocated to the sector.[16] Of course, the moderately well-off to rich peasant cultivators who received land under the land reform programme of the 1960s also benefited from the provision of subsidized new inputs and credits by the government.[17] The favourable output response within this group, which formed the mainstay of Iranian agriculture, accounted for the major part of output growth and marketed surplus in this period (see Karshenas 1990*b*, ch. 6).

Unlike Taiwan, where fixed capital investment played a crucial role in starting off a chain of land-augmenting technical progress by increasing the efficiency of the new seed-fertilizer technology, fixed capital investment in Iranian agriculture was directed towards highly mechanized extensive farming. Land under cultivation grew by more than 40 per cent between 1960 and 1974 as a result of government investment in land infrastructure, which was generally allocated to such mechanized extensive farming. The outcome was that, on the one hand, 60 per cent of agricultural households, composed of poor peasant farmers and landless labourers, existed with extremely low productivity and living standards, and, on the other hand, huge sums were spent on highly capital-intensive mechanized farms which contributed relatively little to growth of employment and output.

Inefficiency of capital investment was not the only source of divergence in the direction and pattern of resource flows in Iran and Taiwan. The other important factor was the rapid growth of consumption in the farm sector in Iran. During the 1963–72 period, real consumption per household in the agricultural sector grew by 2.3 per cent per year, more or less in line with growth of labour productivity in the sector. Between 1972 and 1977, however, consumption per household grew by more than 7 per cent per annum in real terms. Such phenomenal rates of growth in the oil boom period were made possible by sizeable increases in labour income of

[16] This is to some extent reflected in the following statistics relating to 1974. Large-scale farms (above 100 ha) occupied about 15% of the farming area, but did not contribute more than 7% of the total output. Their use of chemical fertilizers was equal to 42.3 kg/ha, compared to a national average of 35.9 kg/ha, and the proportion of land tilled with tractors on these farms was 82.9%, compared with the national average of 56.3 per cent. For more details, see Karshenas (1990b, ch. 5).

[17] Most of the literature on Iranian agriculture justifiably highlights the shortage of funds and personnel which apparently plagued the rural co-operatives, as well as their organizational shortcomings and deficiencies with regard to the size and duration of their loans – which is justified in view of the poor performance of the alternative uses to which the government's agricultural expenditure was put. Nevertheless, one should not underrate the considerable quantity of credit which reached the peasantry through the co-operative network. This could only be appreciated in a historical perspective. The membership of rural co-operatives increased from 750,000 in 1963 to 2m. in 1972 and 3m. in 1977. Just over 70% of credits by specialized banks were distributed to peasant farmers through the co-operatives. In a sample of 339 villages surveyed by the Plan Organization in 1975, 30% of peasants obtained co-operative loans, while 32% used the informal credit market. This should be compared with the figures for 1960, when, according to the Ministry of Internal Affairs (1960, p. 59) only 4% of total agricultural holdings received loans from government sources.

the farm households from non-agricultural activities, large subsidies received by agriculture, and substantial income gains through terms of trade improvements. Unlike Taiwan, where a large part of the income gains in the agricultural sector was siphoned off by the government in direct taxation of the agricultural sector, direct taxation of agricultural incomes as well as land taxes were virtually non-existent in Iran during the study period. With the low degree of financial development in the Iranian countryside, a large part of the new incomes transferred to the sector found its way into consumer expenditure. This was a major source of the net resource flow into agriculture; on the one hand, it increased the inflow of consumption goods from the non-agricultural sector, and, on the other hand, it caused a drop in the growth of agricultural marketed surplus.[18] The drop in the rate of growth of marketed surplus, at a time when food consumption in the economy was rocketing, led to a rapid increase in the food import ratio (see Table 7.3 and Figure 7.2). It is this increase in the food import ratio which has been wrongly attributed to the poor performance of agriculture in the literature (see e.g. McLachlan, 1988). As shown in Table 7.3, the trend rate of growth of value added was between 5.5 per cent and 4.3 per cent respectively for the 1963–70 and 1970–7 periods – respectable growth rates by any standards, and well above the population growth rate.

The central issue with regard to the question of intersectoral resource flows was twofold, therefore. On the one hand, the increase in output and marketed surplus did not seem to be commensurate with the rate of increase of input use in the sector. This was reflected, for example, in the deceleration in the rate of growth of output

Table 7.3 Urban demand, marketed surplus, and imports of food, Iran, 1963–1977 (bn. rials, 1963 prices)

	1963	1970	1977	Trend growth rates	
				1963–70	1970–7
Real food consumption by non-agricultural sector	74.2	130.7	238.2	8.3	9.3
Real food imports	2.0	2.9	32.1	1.9	28.0
Import ratio, (row 2 : row 1) (%)	2.7	2.2	13.5	—	—
Real marketed surplus of food	51.9	86.3	120.6	8.3	6.1
Real value added of agricultural sector	98.4	140.2	177.3	5.5	4.3

Source: Karshenas (1990*b*).

[18] Unfortunately, accurate data on the food consumption of agricultural households are not available. Using the data on average household consumption of food in rural areas provided in the household budget surveys, we estimated trend rates of growth of 1.1% and 5.2% for agricultural households' real consumption of food in rural areas for the 1960–70 and 1970–7 periods respectively. The erratic behaviour of the estimates in some years, and the negative growth of per capita food consumption over the 1960s, which is not plausible, shed doubt on the accuracy of these data. The rapid increase in the trend growth rates is, however, too pronounced to be doubted.

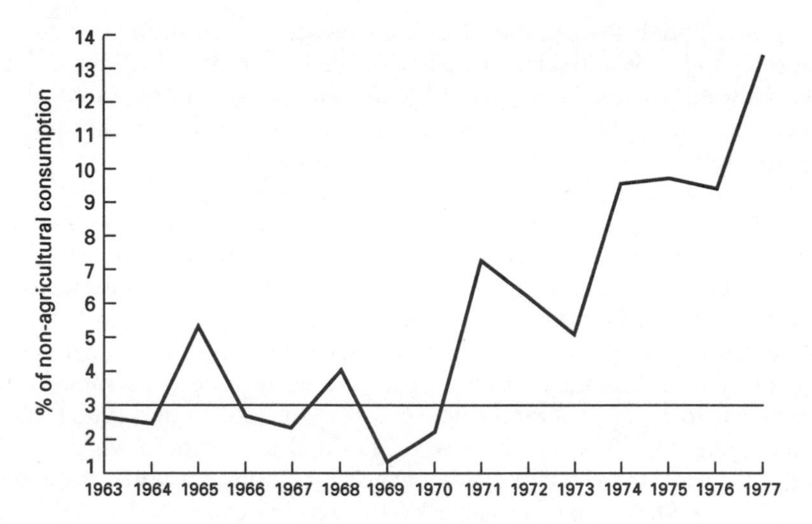

FIG. 7.2 Food import ratio, Iran, 1963–1977

during the 1970s, despite the substantial increase in the inflow of financial and real resources to the sector in that period (see Table 7.3). The second issue was the high rates of growth of consumption of the farm households, particularly in the 1970s, which were not commensurate with the rate of labour productivity growth in the sector. The combination of these two factors led to a situation in which the agricultural sector posed a growing financial burden for the rest of the economy.

7.5 Conclusions

In this chapter we examined the patterns and processes of intersectoral resource flows for Iran during the 1963–77 period. The chapter provided new estimates of intersectoral resource flows from the financial side which indicate that, contrary to common belief, the agricultural sector received a sizeable net financial inflow of increasing magnitude throughout the period under investigation. On the visible side, this consisted of large inflows of capital through government investment and bank credits to the private sector, as well as sizeable factor income inflows, particularly in the form of labour income of the members of farm households engaged in non-agricultural activities. On the invisible side, also, it was shown that, working on plausible assumptions about the base point price system, the agricultural sector seems to have received a substantial inflow of resources through the terms of trade effect. Though the chapter does not claim to have provided exact estimates of the magnitude of resource flows from the financial side, due care has been taken to adopt estimation procedures which err on the conservative side in estimating the inflow of resources to the agricultural sector. More exact estimates have to await further research on calculating resource flows from the real side, which would make it possible to double check the results obtained here. What seems to be beyond dispute, however, is the increasing financial burden of agriculture on the rest of the economy throughout the study period, and the substantial magnitude of net resource flows into the sector during the 1970s.

The chapter also examined the available evidence on the intersectoral resource flows from the real side, in order to identify the major underlying factors which could help to explain the observed pattern of net surplus transfer. First, it was noted that output and labour productivity growth in the agricultural sector were not commensurate with the substantial increase in the new capital and land resources in the sector. This is expected to have an augmenting effect on the net resource inflows by reducing the potential marketed surplus – given the consumption of the farm households – per unit of new producer goods input inflows into the sector. Secondly, during the period under study, the farm sector exhibited consumption growth rates well above those warranted by the rate of growth of labour productivity in the sector, which implied a double squeeze on net surplus outflow by both reducing the rate of growth of marketed surplus and increasing the demand for consumer goods purchases from other sectors of the economy. This resulted from the fast rates of growth of farm household incomes, mainly as a consequence of the income gains through the terms of trade effect and of the growth of factor income inflow into the sector. What distinguishes the experience of Iran from that of Japan and Taiwan is the absence of any attempt by the government to siphon off part of this new income by direct taxation of the farm sector. The evidence suggests that, with the low degree of financial development in the Iranian countryside, the new income flows were largely directed towards increased consumption of the farm households.

8

Intersectoral Resource Flows in Japan, 1888–1937

8.1 Introduction

In this chapter we examine the pattern of intersectoral resource flows in the early process of development during the 1888–1937 period in Japan. There are various studies of intersectoral resource flows in Japan for this period (see e.g. Ishikawa, 1967*a*, 1988; Ohkawa, Shimizu, and Takamatsu, 1978, 1982; Mundle and Ohkawa, 1979; Teranishi, 1982). We shall draw on the findings of all these studies in this chapter, but the main results which form the central theme of our discussion will be those of Ohkawa *et al.* (1982). The chapter is organized in the following way. In the next section we discuss the pattern of intersectoral resource flows from the real side, where two subperiods are distinguished: the first one roughly spans the period from the Meiji restoration (1868) to the First World War; and the second covers the interwar period. In Section 8.3 we examine the relation between the changing pattern of intersectoral resource flows and agricultural productivity growth during these two subperiods. The financial mechanisms of surplus flow are discussed in Section 8.4, and a summary and the main conclusions of the chapter are given in Section 8.5.

8.2 Resource Flows on the Real Side

The measurement of intersectoral resource flows from the real side in the case of Japan has been handicapped by the shortcomings of the available historical data. Various researchers, in presenting their estimation results, have been cautious to point out that the calculations should be regarded as tentative and, at best, indicative of the direction of resource flows and their long-term trends rather than as exact estimates. Successive research has produced different and improved results over time. The estimates discussed in this chapter are based on the latest of such studies, by Ohkawa *et al.* (1982), which provides the most up-to-date and comprehensive figures. Table 8.1 shows the estimates by Ohkawa *et al.* of different components of intersectoral resource flows for the 1888–1937 period at current prices. These estimates are based on the sectoral distinction between agriculture and non-agriculture rather than on the institutional one between farm and non-farm households.[1] These estimates are also reproduced in terms of ratios of agricultural value added in Table 8.2.[2]

As can be seen, agricultural marketed surplus increased from about 54 per cent of value added at the beginning of the period to about 85 per cent by 1937. The purchases of the agricultural sector, on the other hand, start from a lower base of about 48 per cent of value added, and, growing at a relatively slower pace – particularly during the 1888–1917 period – remain consistently below the marketed surplus. As a consequence, there seems to have been an outflow of resources from the agricultural sector throughout the period, with the exception of the depression years of 1929–32. During this period, both agricultural sales and purchases at current prices registered a substantial decline, with sales dropping proportionately much faster, due largely to a sharp fall in the agricultural terms of trade. Since this was the result of special circumstances, largely related to the collapse of the silk industry and the advent of the world depression, and as intersectoral resource flows returned to their earlier pattern with the end of the depression, we may maintain that, during the period as a whole, the net product contribution of agriculture at current prices was positive.

A notable feature of the pattern of intersectoral resource flows during the period under study is the acceleration in the rate of surplus outflow from agriculture in the

Table 8.1 Value of intersectoral commodity flows, Japan, 1888–1937 (¥m. per year, current prices)

Year	Value of agricultural sales (X)	Value of agricultural purchases				Net balance (X−M)
		Total (M)	Consumer goods	Producer goods	Investment goods	
1888–92	217	191	61	102	28	26
1893–7	310	267	81	145	41	43
1898–1902	486	407	108	243	56	79
1903–7	712	505	144	292	69	207
1908–12	874	684	190	409	85	190
1913–17	1182	891	249	542	100	291
1918–22	2761	2453	567	1661	225	308
1923–7	2810	2701	569	1876	256	109
1928–32	2014	2079	454	1411	214	−65
1933–7	2363	2211	505	1499	207	152

Source: Ohkawa, Shimizu, and Takamatsu (1982).

[1] In providing these estimates, Ohkawa *et al.* (1982) divided farm household investment and consumption into its agricultural/non-agricultural components. This practice is not followed in the other country studies discussed in this book, and hence, in making cross-country comparisons, it is important to keep this point in mind.

[2] Agricultural value added here refers to the national accounts' estimates of gross value added in agriculture plus that part of investment in agriculture which is produced by agricultural households, as measured by Ohkawa *et al.* (1982).

Table 8.2 Composition of intersectoral commodity flows, Japan, 1888–1937 (% share of agricultural income)

Year	Value of agricultural sales (X)	Value of agricultural purchases				Net balance (X−M)
		Total (M)	Consumer goods	Producer goods	Investment goods	
1888–92	54.3	47.8	15.3	25.5	7.0	6.5
1893–7	56.1	48.3	14.6	26.2	7.4	7.8
1898–1902	59.3	49.7	13.2	29.7	6.8	9.6
1903–7	67.3	47.7	13.6	27.6	6.5	19.6
1908–12	66.2	51.8	14.4	31.0	6.4	14.4
1913–17	75.5	56.9	15.9	34.6	6.4	18.6
1918–22	77.3	68.7	15.9	46.5	6.3	8.6
1923–7	81.6	78.4	16.5	54.5	7.4	3.2
1928–32	83.8	86.5	18.9	58.7	8.9	−2.7
1933–7	84.9	79.5	18.2	53.9	7.4	5.5

Source: Ohkawa, Shimizu, and Takamatsu (1982).

first half of the period, and its deceleration during the second half. The rate of net surplus outflow increased from 6.5 per cent of the agricultural sector's value added in 1888–92 to more than 18.5 per cent in 1913–17, but from then on it followed a declining trend, culminating in the value of 5.5 per cent in 1933–7 (see Table 8.2). This cyclical behaviour of surplus outflow was largely due to the changing behaviour of the agricultural purchase ratio during the two subperiods. During the first period, purchases of intermediate and investment goods by agriculture as a ratio of agricultural income remained remarkably stable, and even followed a slightly downward path. The increase in the rate of consumer goods purchases in this period was also moderate. However, in the second half of the period, that is, from 1917 onwards, the rate of agricultural purchases as a proportion of value added accelerated rapidly. The cyclical behaviour of intersectoral resource flows demarcates two distinct phases of development of Japanese agriculture, to which we shall return shortly.

The estimates of surplus outflow from agriculture in Tables 8.1 and 8.2 are valued at current prices. To get a measure of real intersectoral resource flows, we need to take into account also the effect of relative price changes, or the terms of trade effect. As Table 8.3 and Figure 8.1 show, the terms of trade moved consistently in favour of agriculture until the 1920s. During the 1920s and the 1930s, Japan developed a growing food shortage and resorted increasingly to food imports from the colonies in order to stabilize the domestic food prices. Despite the growing food imports and, in particular, the sharp dip in the terms of trade during the depression years, agricultural terms of trade were more than 33 per cent higher than they were at the beginning of the reference period. It should be noted that, despite the growth in food imports (notably rice) and the reversal of agricultural

Table 8.3 Terms of trade of agriculture, Japan, 1888–1937

Year	Price indices		Net barter terms of trade	Income terms of trade
	Agricultural sales (P_a)	Agricultural purchases (P_n)		
1888–92	199.00	100.00	100.00	0.00
1893–7	129.31	115.36	112.10	25.00
1898–1902	164.03	131.75	124.50	60.80
1903–7	195.45	151.87	128.70	74.20
1908–12	213.24	157.26	135.60	114.20
1913–17	228.26	184.08	124.00	93.70
1918–22	482.02	324.37	148.60	247.30
1923–7	452.70	305.88	148.00	286.40
1928–32	297.96	252.08	118.20	127.00
1933–7	331.75	248.13	133.70	224.60

[a] ¥m., at 1988–92 prices.

Source: Ohkawa, Shimizu, and Takamatsu (1982).

terms of trade improvements in the 1920s and the 1930s, the degree of protection afforded to Japanese agriculture was accelerating rapidly during these two decades. After the rice riots and the sharp price increases in 1918, it became declared government policy to achieve rice self-sufficiency and price stability through the development of agriculture both at home and in the colonies (Anderson, 1983). This was achieved by the late 1920s, largely through the rapid productivity growth in the two colonies of Korea and Taiwan. During this period, however, the agricultural market of the empire as a whole was highly protected, and the relatively lower cost of production in the colonies, as noted in the case of

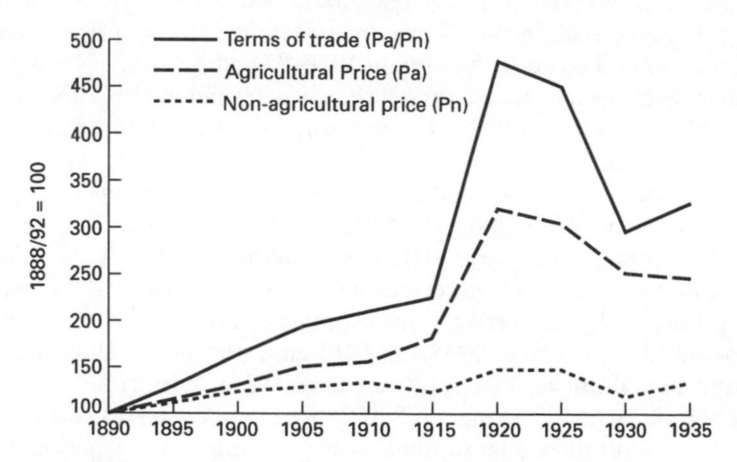

FIG. 8.1 Agricultural terms of trade, Japan, 1888/92–1933/7

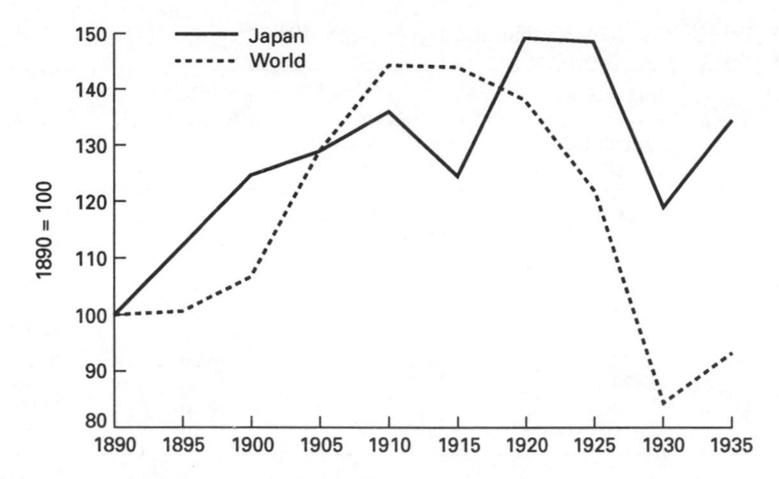

FIG. 8.2 Agricultural terms of trade, Japan and the world, 1890–1935

Taiwan in the previous chapter, was translated into high rents and government exactions from agriculture. According to estimates by Anderson (1983), the nominal protection of rice, which stood at about 16 per cent in 1903–7 and 20 per cent in 1918–22, increased to about 60 per cent by the late 1930s. This is also clearly demonstrated in Figure 8.2, which shows the trends in the agricultural terms of trade in Japan and in the international economy over the 1900–35 period. As can be seen, relative to the international levels, agricultural terms of trade in Japan improved considerably from 1915 onwards.

The terms of trade improvements in favour of the agricultural sector during the reference period meant relatively large real income gains for the farm households. As Table 8.3 shows, such income gains, measured in 1888–92 prices, increased from ¥25m. in 1893–7 to about ¥247m. in 1918–22 and ¥224m. in 1933–7. Such substantial invisible income transfers could easily overshadow the visible transfers shown in Table 8.1, and hence the real intersectoral resource flow measure could be very different from the current value measure. To estimate the real net intersectoral resource transfers, however, we need first to choose an appropriate base year for evaluation. Given the erratic movements in the agricultural terms of trade, the results would naturally be sensitive to the choice of the base year. Table 8.4 shows different estimates of intersectoral resource flows at 1895, 1915, 1925, and 1935 base-year prices, alongside current value estimates. As can be seen, real surplus outflow estimates, measured at 1895 and 1915 base year prices, show a growing outflow from agriculture until 1917, turning to an inflow during the 1920s and the 1930s. On the other hand, taking 1925 or 1930 as the base year, we observe a continuous surplus outflow on a substantial scale throughout the period. Choosing an appropriate base year on theoretical grounds, e.g. proximity to world prices or to some notion of an equilibrium situation, poses its own problems, as discussed in Chapter 2. For example, it may be argued that the two earlier base years are more appropriate because, during the 1920s and the 1930s, the Japanese market was so highly protected, with prices much higher than world prices, and, furthermore, it was highly affected by rice imports from the colonies, with artificially depressed wage and consumption levels and with costs of production of little relevance to those prevailing in Japanese agriculture. However, as can be observed from Figure 8.2, world prices seemed to be extremely volatile during the reference period, and hence the notion of using them also becomes problematic. This is a clear example of the problems involved in trying to determine a 'correct' measure of the absolute value of real surplus transfer, as discussed in Chapter 2. However, it is crucial to note that, despite the differences in absolute values between the various measures of surplus flow presented in Table 8.4, they all share one common characteristic. In all the cases, whether measured at current prices or real values, there is a clear upward trend in the net outflow of resources from agriculture between 1888 and 1917, which is reversed in the subsequent period and, in some cases, even turned into an inflow.

To recapitulate: the evidence on intersectoral resource flows shows a net surplus flow out of the agricultural sector during the 1888–1917 period, whether measured at current prices or constant values with different base years. Furthermore, during this early period, one can observe an increasing trend in the net outflow of resources from the agricultural sector. Barring the questions which may arise as to the reliability of the data, it may be concluded that, during this early period, the net product contribution of agriculture to the Japanese economy was positive and increasing. During the 1920s and the 1930s, the picture with regard to the direction of net resource flows is more ambiguous. At current prices and valued at 1925 and 1935 prices, the positive net surplus outflow continues for the rest of the period as

Table 8.4 Net intersectoral commodity flows, Japan, 1888–1937

Year	Current prices (X−M)	Net commodity flows at base year prices (¥m.)			
		1895	1915	1925	1935
1888–92	26	60.3	143.7	398.1	246.0
1893–7	43	43.0	121.1	377.3	221.0
1898–1902	79	26.8	107.7	396.4	216.4
1903–7	207	87.4	219.4	632.0	383.4
1908–12	190	28.2	134.9	525.0	280.5
1913–17	291	111.2	291.0	863.7	516.9
1918–22	308	−131.7	−84.6	279.9	23.8
1923–7	109	−216.0	−208.6	109.0	−131.8
1928–32	−65	−77.4	24.7	537.2	196.0
1933–7	152	−106.9	−14.4	498.9	152.0

Source: Tables 8.1 and 8.3.

well. However, at 1895 or 1915 prices, the direction of surplus flow changes into a net inflow of resources into agriculture. In all cases, however, there is an unambiguous cyclical movement in the trend of surplus flow – i.e. an increasing outflow during the early period which is reversed or even turns negative during the 1920s and 1930s. It is this changing behaviour in surplus flow over time which produces the key to understanding the relation between agricultural growth and intersectoral resource processes, which will be the subject of the next two sections of this chapter.

8.3 Agricultural Growth and Resource Flows

As already observed in previous chapters, output and productivity growth in agriculture appear to be the key factors in determining the potential for surplus outflow from the agricultural sector. Labour productivity in Japanese agriculture, even at the inception of the observation period, was well above that of the other countries in the sample, as noted in the discussion of the initial conditions in Chapter 4. This was the major underlying factor which explained the high rates of marketed surplus in Japanese agriculture. Growing levels of labour productivity, and hence income, implied a high degree of diversification in consumer demand in the farm sector and a proportionately lower rate of self-consumption of agricultural produce. Similarly, the rate of growth of marketed surplus is expected to be directly related to the rate of growth of labour productivity.[3] The rate of marketed surplus, starting from about 54 per cent of agricultural value added at the beginning of the observation period, increased to about 90 per cent by the end of the 1930s. This rapid increase was a reflection of the fast rates of labour productivity growth during the period as a whole. Of course, the income gains arising from labour productivity growth and other sources of income can lead to a countervailing increase in the purchases of consumer goods by farm households from other sectors. This aspect of the resource flow mechanism is closely connected to the financial side of surplus transfer which will be discussed in the next section. At this stage we can maintain that labour productivity growth is a key factor in the increase in the *potential* for surplus outflow from agriculture. This statement, however, has to be qualified further, as the positive effects of labour productivity growth on marketed surplus may be neutralized by a disproportionate increase in the purchased inputs and investment goods which are necessary to bring it about.

[3] Surely, other factors such as the nature of agrarian institutions, organization of production, farm size, degree of utilization of wage labour, and commercialization of agriculture in general also play a crucial role in determining the marketed surplus of the agricultural sector. However, during the period under study, no major institutional or organizational changes took place in Japanese agriculture, and farm size and degree of utilization of wage labour also seemed to be stable (Ohkawa and Rosovsky, 1960). The growth of marketed surplus was thus highly correlated with the growth of labour productivity in the sector.

A further set of factors which affect the potential outflow of net surplus from agriculture, therefore, relates to the efficiency of resource use and the nature of technological progress in agriculture. We will turn now to this set of issues.

The productivity of labour more than doubled during the 1880–1935 period, growing at an average annual rate of 1.8 per cent (see Table 8.5). This was partly due to the decline in the agricultural labour force and the increase in the land/labour ratio during the reference period. The agricultural labour force declined by about 2.1m., or 13 per cent, between 1880 and 1935.[4] The fall in the agricultural labour force was due to a combination of the slow rate of population growth and the high rate of absorption of agricultural labour in other sectors of the economy.[5] During the same period, the area of arable land increased from 4.7m. hectares to 6.1m., or by 24 per cent. The result was that the land/labour ratio increased by about 45 per cent during the 1880–1935 period (see Table 8.5), thus distinguishing the experience of Japan in this regard from other countries in our sample, with the exception of Iran. The growth of labour productivity in Japan, however, is only partially explained by the increase in the land/labour ratio.

Table 8.5 Output, input, and productivities in Japanese agriculture, 1880–1935

Year	Total output	Total input	Productivities			Land/labour ratio
			Total	Labour	Land	
1878–82	100.0	100.0	100.0	100.0	100.0	100.0
1883–7	110.6	101.3	109.2	110.5	108.6	101.8
1888–92	119.7	102.8	116.4	119.3	115.1	103.7
1893–7	122.5	105.4	116.2	121.8	115.2	105.7
1898–1902	138.6	108.2	128.1	136.4	126.5	107.8
1903–7	152.1	111.5	136.4	148.8	134.7	110.5
1908–12	170.4	117.8	144.7	167.4	143.4	116.7
1913–17	193.3	119.9	161.2	196.8	157.0	125.4
1918–22	205.5	119.3	172.3	228.8	162.6	140.7
1923–7	208.2	121.6	171.2	239.9	166.6	144.0
1928–32	223.4	127.3	175.5	249.3	176.3	141.4
1933–7	236.3	128.7	183.6	267.3	183.3	145.8
Annual growth rates						
1880–1920	1.8	0.4	1.4	2.1	1.2	0.9
1920–35	0.9	0.5	0.4	1.0	0.8	0.2
1880–1935	1.6	0.5	1.1	1.8	1.1	0.7

Source: Yamada and Hayami (1979*a*).

[4] The agricultural labour force grew by an annual average rate of 0.1% over the 1880–1900 period, and by −0.6 and −0.1% over the 1900–20 and 1920–35 periods respectively (Ohkawa and Shinohara, 1979, table A18, p. 293).

[5] Population growth rates rose from about 1% in the late 1880s to 1.5% in the 1930s. The share of non-agricultural labour force in the total increased from 29% in 1888 to 57% in 1937 (Ohkawa and Shinohara, 1979).

The major part of the labour productivity increase (about 60 per cent) over the period as a whole is explained by the increase in yields, which signifies the importance of the land-saving technological innovations for agricultural development in Japan during this period.

In relation to the output and productivity growth in agriculture, two subperiods can be distinguished which also broadly correspond to the periods of changing trends in surplus transfer discussed in the previous section. The first subperiod, which really runs from the Meiji restoration to the First World War, was one of relatively high productivity and output growth in agriculture. Agricultural output outstripped population growth in this period, and, in addition to providing food and raw materials for the fast-expanding non-agricultural sector, agriculture also produced an export surplus which made an important contribution to industrial development (Ohkawa and Rosovsky, 1960). Productivity growth in agriculture at this time was based on a constant and incremental improvement in technological practices within the traditional agrarian institutions inherited from the Tokugawa period. Small family farms with, on average, about one hectare of land per household, remained the main operational units of production, both on the owner-occupied and rented land. Technological innovations took the form of land improvement and extension of irrigation, as well as the introduction of new seeds, fertilizers, and better methods of cultivation. The government and the rural-based landlord class played an important role in these technological improvements by both providing the finance for bulkier land-improvement investment projects and helping to promote the technological innovations (see Ohkawa and Rosovsky, 1960). According to the estimates by Yamada and Hayami (1979*a*), about 75 per cent of the output growth in this period is explained by productivity growth and only 25 per cent by the increase in inputs (see Table 8.5). This indicates the fast rates of technological innovation and the high and improving degree of efficiency of resource use to be found in Japanese agriculture. This was particularly evident in the output response to the application of new inputs purchased from outside the agricultural sector, as witnessed by the low, and even slightly declining, ratio of purchased inputs to value added reported in Table 8.2 above.

The second subperiod, namely, the interwar period, was one of relative stagnation in output and productivity. During this period the non-agricultural sectors of the Japanese economy continued their fast rates of growth, and a widening gap developed between the demand and supply of agricultural products, which, as we have already noted, was covered by increasing imports from the colonies. The sluggish agricultural productivity growth also implied an increasing degree of protection of the agricultural sector by the government in an attempt to maintain the incomes of the farm households and landlords, who formed a strong political lobby during this period (Anderson, 1983). The slow-down in the rate of growth of output over this period was partly due to the relatively smaller increase in inputs such as land and other fixed and working capital inputs.[6] To a larger extent, however, it was due to the slow-down in the rate of agricultural productivity. The contribution of total factor productivity, which stood at 75 per cent of the output growth during the 1880–1920 period, declined to 44 per cent during the 1920–35 period, while the contribution from the rise in inputs increased from 25 per cent to 56 per cent. This was also reflected in the rapid increase in the rate of purchased intermediate and capital goods/value added ratios reported in Table 8.2 and discussed in Section 8.2 above. In fact, it would be plausible to assume that the slow-down in the rate of increase of agricultural inputs was itself caused by the slow-down in the rate of productivity growth which, *ceteris paribus*, implied a lower rate of return on investment in the agricultural sector than in the earlier subperiod. To investigate the causes of the slow-down in the rate of productivity growth in the interwar period would take us far afield. According to Ohkawa and Rosovsky (1960), by the end of the First World War the main sources of technological innovation within the traditional institutional set-up of Japanese agriculture seemed to have been exhausted. Further technological progress required radical institutional reform, which was hindered by social and political obstacles.[7] Furthermore, the institutional changes which did take place were not conducive to technological innovation. An important example of this was the increase in parasitic or absentee land-ownership, and the gradual fading-away of the Japanese landlord-entrepreneur who had played a dynamic role in the introduction of new innovations in the earlier period. According to Ohkawa and Rosovsky (1960), during the interwar period, 'landlord interest was shifting from production to marketing, and their collective efforts came to be concentrated on maintaining the price of rice' (ibid., p. 59).

Clearly, the change in the pace of output growth, and particularly the slow-down in productivity growth between the two subperiods, had a direct bearing on the potential of agriculture to generate a surplus for transfer to other sectors. The mechanisms through which the surplus transfer took place, and the degree to which this potential was realized, are the topic of the next section.

[6] For example, the rate of increase in cultivated land declined from 0.7% per annum in 1900–20 to 0.1% per annum in 1920–35. During these respective periods the annual rates of growth of real fixed capital declined from 1.3% to 0.9%, and that of current inputs from 4.7% to 3.2% (Yamada and Hayami, 1979*a*).

[7] To quote Ohkawa and Rosovsky (p. 59): 'the Japanese farmer, given the prevailing system of cultivation, had reached his most efficient method of production in the teens of the twentieth century, and now he was not able to make further impressive gains. ... The entire traditional agricultural complex which had served Japan quite well since the early changes of the Tokugawa Era, and which had been spectacularly successful during the Meiji and part of Talsho, now entered a far less brilliant period. Perhaps the greatest problem lay in the fact that major changes were politically, socially, and culturally quite impossible.'

8.4 Financial Mechanisms of Surplus Flow

The contribution of different financial channels to the resource flow process in Japan for the period 1918–22 is shown in Table 8.6. Though there were some important changes in the pattern of financial flows over time which will be discussed below, the figures for the 1918–22 period will serve to highlight the salient features of financial flows for the study period as a whole. The pattern of financial flows at this time was largely shaped by the prevailing economic institutions, particularly in the agricultural sector, and the fast rates of growth and structural change in the economy as a whole.

One feature of the financial flows in Japan which immediately stands out from the other countries in our sample, particularly India and colonial Taiwan, is the substantial inflow of factor income to the agricultural sector. This was a consequence of the combination of low factor outflow through rents and large inflow in the form of labour income from the non-agricultural activities of farm household members. The former was due to the peculiarity of the agrarian relations in Japan, where, unlike colonial Taiwan, landlords were mainly rural-based farmers, and their rent income did not constitute a financial outflow from agriculture. The second item, namely, the wage income of agricultural households from the non-agricultural sector, was also considerably larger, relative to other financial flows or the size of agricultural income, than that of the other countries in the sample. The importance of this source of financial inflow in the case of Japanese agriculture stemmed from the rapid growth of the non-agricultural sector and the fast pace of structural change in the Japanese economy over this period. This type of wage income inflow from the non-agricultural activities of the farm households could be regarded as part of the contribution of farm households to economic growth through labour transfer, as discussed in Chapter 1. It has been argued in the literature that, given that this wage income was supplementary to farm household income, it had a particularly important impact on industrial accumulation by keeping industrial wages low (see Shinohara, 1970).[8]

The current official transfers have received much attention in the literature on the financial contribution of agriculture to the early economic development of Japan, particularly during the Meiji period. As can be seen from Table 8.6, net current official transfers remained a significant source of resource outflow from agriculture, constituting about 70 per cent of total net resource outflow, even during the 1918–22 period. However, the contribution of this source to total net resource outflow, both relative to agricultural income and as a share of non-agricultural investment, was declining continuously throughout the period under study. The contribution of land taxes to public sector finances was indeed

[8] Apart from a 'part-time' labour contribution, as a supplementary source of farm household income, the agricultural sector also made a 'permanent labour' contribution to industrialization by providing a constant supply of migrant labour to industry. It is estimated that, during the 1888–1937 period, more than 70% of the increase in labour force in non-agricultural industries came from the agricultural sector (Umemura, 1979).

Table 8.6 Financing of net agricultural resource outflow, Japan, 1918–1922 (annual averages, ¥m., prices)

	1918–22	
Net resource outflow (R)	308	(9)
(*a*) Agriculture's sales (X)	2761	(77)
(*b*) Agriculture's purchases (M)	2453	)69)
Financing items		
1. Net outflow of factor income[a] ($F_a - Y_f$)	−1139	(−31)
(*a*) Land rents[b]	133	(4)
(*b*) Labour income	−1272	(−35)
3. Net outflow of current transfers[c] ($T_{fg} - T_{gf}$)	134	(4)
(*a*) Taxes	288	(8)
(*b*) Subsidies	−75	(−2)
4. Net outflow of capital transfers	225	(6)
(*a*) Net Private[d] ($K_{fo} - K_{of}$)	247	(7)
(*b*) Government investment ($K_{fg} - K_{gf}$)	−22	(−1)
5. Notional consumption transfer[e]	1088	(30)

Note: Figures in brackets are % share of agricultural income.

[a] Excludes interest payments to non-farming households.

[b] Rents to non-farming or absentee landlords.

[c] Refers to government's net transfers only.

[d] Includes the increase in cash in circulation in the farm sector.

[e] This is a residual item, mainly composed of the notional consumption transfers from the agricultural sector to the farm households calculated by Ohkawa et al. (1982). It also includes current private transfers.

Source: Ishikawa (1988), and Ohkawa, Shimizu, and Takamatsu (1982).

substantial during the period preceding the First World War. Land taxes formed about 80–90 per cent of total central government revenue during the last two decades of the nineteenth century, and, though gradually declining from the turn of the present century, on the eve of the First World War they still constituted about 40 per cent of total revenue. In the interwar period the significance of land taxes declined rapidly, and by the end of the 1930s they formed no more than about 10 per cent of total government revenue (see Table 8.7). In this latter period, income taxes and business taxes, largely financed from the non-agricultural sector, replaced land taxes as the main source of government revenue.

A notable aspect of the Japanese taxation policy in the period under study, which was similar to that of Taiwan but very unlike that of India and Iran, was the much higher burden of direct taxation on agriculture compared to non-agriculture (see Table 8.7). Despite the narrowing of the gap between the relative tax burdens over the period, by the end of the 1930s the burden on agriculture was still twice as high as on the non-agricultural sector. Nevertheless, given the relative stagnation of the agricultural sector and the fast rates of growth of the industrial and services sectors in the interwar period, the contribution of agriculture to total tax revenue was decreasing steadily, and by the end of the period the non-agricultural sector

Table 8.7 Composition of central government tax revenues and burden of taxation, Japan, 1890–1935

Year[a]	% share in government tax revenue				Tax burden[b]	
	Land tax	Income tax	Business tax	Customs duties	Agriculture	Non-agriculture
1890	85.6	2.4	1.6	10.4	15.5	2.3
1895	80.4	3.3	2.8	13.5	12.4	2.0
1900	63.2	7.8	8.6	20.4	12.1	3.2
1905	55.8	15.5	12.3	26.4	11.2	5.4
1910	42.9	18.3	13.5	25.3	12.5	6.4
1915	37.6	26.0	12.9	23.5	12.9	4.5
1920	18.3	47.4	14.2	21.1	9.2	5.4
1925	15.5	45.0	12.8	26.7	10.5	5.2
1930	15.8	42.9	11.7	29.6	9.7	4.3
1935	10.7	49.4	11.5	28.4	7.8	4.2

[a] Figures refer to 5-year averages centred on the year shown.
[b] Direct taxes collected as % of income produced in agriculture and non-agriculture.
Source: Based on Ohkawa and Rosovsky (1960, tables 14 and 15).

supplied the major share of revenues.[9] In fact, given the decline in the tax burden and the sluggish growth of the agricultural sector, the absolute value of the direct tax revenue from the agricultural sector fell from its peak of ¥310m. in 1923–7 to about ¥195m. during the 1930s (Ohkawa *et al.*, 1982).

To consider the net contribution of agricultural surplus on the official account, in addition to taxes, one should also take into account the current and capital transfers from the government to the farm sector. It appears that the inflow of government finance through current subsidies and capital investment accelerated during the interwar period, as tax revenues from the sector had just peaked and then begun to decline. Government investment in agriculture at current prices, which had increased from about ¥1m. to about ¥7m. between 1888 and 1917, leapt to about ¥60m. in the mid-1920s and to more than ¥75m. by the end of the 1930s (Ohkawa *et al.*, 1982, p. 38). A similar pattern could be observed in relation to government subsidies to agriculture and other current transfers.[10] As a consequence, net surplus outflow on the official account, which formed an important source of overall net surplus outflow in the early period, declined continuously

[9] Note that the share of agriculture in total net domestic product declined from 46% in 1890 to 31% in 1910 and 18% in 1937 (Ohkawa and Shinohara, 1979, table A10).

[10] For example, subsidies to the agricultural sector, which remained at zero until the First World War, increased rapidly during the 1920s and the 1930s, and by the end of the period absorbed 20% of total subsidies granted by the government (see Ohkawa and Rosovsky, 1960). Other current transfers mainly aimed at maintaining farm household incomes during the agricultural crisis of the 1920s and 1930s also increased rapidly in the latter period (see Ohkawa *et al.*, 1982, p. 38).

over time, and, indeed, turned negative in the late 1920s and in the 1930s.[11] It appears, therefore, that, in line with the popular view during the Meiji era, the government's budgetary policies played an important role in surplus outflow from the agricultural sector – though in the interwar period this process seems to have been reversed, with the budgetary mechanism turning into a positive source of surplus flow into agriculture.

Another important source of surplus outflow from agriculture in Table 8.6 takes the form of voluntary savings by farm households channelled into the non-agricultural sector through the credit mechanism. As can be seen from the table, net private capital transfers constituted more than 80 per cent of the total net financial outflows from agriculture during the 1918–22 period. However, as with the official flows, net private capital transfers underwent significant changes during the period under study, though in the reverse direction. In contrast to the net official transfers, the private capital flows seem to have been negligible, even forming an inflow into agriculture during the early Meiji era. This turned into a positive outflow after the First World War and increased in magnitude throughout the interwar period.[12] This process was related to a number of factors. The high, and increasing, net inflow of factor income to the farm sector, combined with the growing net official transfers and subsidies to agriculture, seems to have contributed to the generation of surplus funds at a time of relatively stagnant productivity in agriculture. This was further reinforced by the slow rate of increase in farm household consumption and increasing savings ratios, which are attributed to the slow change in consumption habits and traditional lifestyle in the rural communities in this period (see Ohkawa and Shinohara, 1979; Ohkawa *et al.*, 1982). Given the sluggish technological change and the low rate of return on agricultural investment in the interwar period, the surplus agricultural funds increasingly found their way into alternative forms of financial investment with much higher rates of return than real investment in agriculture (Ishikawa, 1988).

The final item in Table 8.6 refers to the notional consumption transfer, which arises from the way in which Ohkawa *et al.*, (1982) measure agricultural consumption as distinct from farm household consumption when estimating net commodity flows on a sectoral basis. The rationale behind this type of accounting, according to Ohkawa *et al.* is that, since wage labour inflow from non-agricultural activities forms a large share of farm household income, to arrive at a measure of

[11] Net official transfers, inclusive of capital investment by government in agriculture, amounted to ¥47m. on average during the 1888–92 period, which was 80% higher than total net surplus outflow from agriculture. It increased to ¥117m. in 1913–17, which was now only 40% of total financial outflow, and became negative from the late 1920s onwards (Ohkawa *et al.*, 1982).

[12] It is interesting to note in this context that rural credit co-operatives played an important role in this process. According to Ishikawa (1988), from 1917 onwards the deposits in credit co-operatives exceeded lending to the farm households, with the deposit surplus growing over time – by 1932, deposits were 4 times as much as loans. The surplus was either deposited in commercial banks or invested in securities with much higher rates of return than could be obtained from investment in agriculture.

consumption by direct agricultural producers, one needs to separate the consumption financed by this additional source of income. Given that consumption activity takes place within the farm household as a unit, this separation always remains rather *ad hoc*. The way Ohkawa *et al.* have undertaken the measurement is to divide total consumption in proportion to the agricultural and non-agricultural sources of farm household income. Thus, to provide financial accounts which are based on institutions, comparable to the sectoral flows measured by Ohkawa *et al.*, we have had to include that part of consumption related to non-agricultural activities in the table as a notional transfer out of the agricultural sector. As can be seen, this is a relatively large item which is more or less equal to the factor income inflow through wage labour income. If we exclude this item from the financial accounts in Table 8.6, we arrive at a figure for net resource flows which is negative and of the same order of magnitude as that estimated from the financial side by Teranishi (1982). Thus, the difference between the estimates of Teranishi and Ohkawa *et al.* lies in their treatment of consumption flows. If one takes total purchases of consumer goods by farm households as an inflow to agriculture in the real-side estimates provided by Ohkawa *et al.*, then the latter measure becomes equivalent to Teranishi's estimates.[13] Given the size of the factor income inflow into the farm sector, different treatments of farm household consumption expenditure would result in substantial changes in the absolute value of net surplus transfer at current prices. However, as long as a consistent treatment is followed for the period as a whole, this is unlikely to change our conclusions on the changing trends of the intersectoral resource flows.[14]

8.5 Conclusions

In this chapter we have discussed the pattern of intersectoral resource flows for Japan during the 1888–1937 period. As we have observed, according to the estimates by Ohkawa *et al.* (1982), there seems to have been a net resource outflow from agriculture throughout the period, when valued at current prices. At constant prices, however, the results differ according to the base year prices chosen for the measurement of the terms of trade effect. In general, the net surplus outflow seems to have been positive and increasing during the 1888–1917 period, and declining during the 1918–37 period, sometimes turning negative in the latter period,

[13] Of course, this is not the only difference between the estimates of these two sources. Their estimates also differ slightly in relation to the current and capital transfers (see Ishikawa, 1988, table 10.1), but the main difference arises from the consumption accounting adopted by Ohkawa *et al.*

[14] The information provided in Ohkawa *et al.* (1982) unfortunately does not allow a re-estimation of their results under alternative assumptions. However, assuming a 30% difference between the farm household and the notional agricultural household consumption (based on data provided in Mundle and Ohkawa, 1979), we arrive at the following results if we attribute total farm household consumption to the agricultural sector: during the 1888–1917 period, net surplus outflow remains positive and increases at current prices, and during the 1918–37 period it turns negative and follows a declining trend.

depending on the base year prices adopted. This changing trend in the pattern of resource flows demarcates two distinct phases in the development of Japanese agriculture.

The first phase was one of rapid technological change and productivity growth in Japanese agriculture. Productivity growth over this period was based on the introduction of a stream of land-saving innovations well suited to the traditional agrarian organizations and small, family-sized farming units. Technological progress in this period took place without the need for any major institutional reorganization of the traditional agriculture and with low capital intensity. The role of the rural-based landlord class and the government, both in the diffusion of the new technological innovations and in infrastructural support, was crucial in the process of technological change. The major part of output and productivity growth in this period was due to technological change, and only a small part (25 per cent) was due to the increase in inputs. As a consequence, the ratio of output to purchased inputs from other sectors (both for current and fixed capital investment) remained stable and even showed a slight increase during this phase of agricultural development.

The second phase was one of agricultural stagnation and low productivity and output growth. During this period, the small output increases which were achieved were mainly based on growth of inputs, and technological change played a minor role in the growth of output. It appears that, during this period, the sources of technological improvement within the traditional institutions of Japanese agriculture were exhausted, and returns to new investment in the sector within the traditional agrarian organizations were diminishing. As a consequence, the rate of increase in inputs compared to the earlier period also slowed down, and hence agricultural output stagnated. The slow-down in the rate of technological innovation in this period was reflected in a rapid decline in the ratio of agricultural output to the inputs purchased from the non-agricultural sector. The stagnation of agricultural output and productivity meant that the role of agriculture as a potential source of surplus became increasingly limited.

The differential in the rates of productivity and output growth between the two subperiods also had important implications for the movement in the terms of trade and other mechanisms of financial resource flow. During the dynamic phase of agricultural growth prior to the First World War, the terms of trade improved only slowly in favour of agriculture, and agricultural protection remained moderate. In fact, during the 1900–15 period, the agricultural terms of trade in Japan relative to world prices declined substantially. This gap was closed rapidly during the inter-war period, however, as the agricultural terms of trade in Japan relative to world prices improved rapidly. With the slow-down in the rate of productivity growth in this latter period, the government had to intervene to protect agricultural incomes, and the degree of protection afforded to agriculture increased substantially.

With regard to current and capital flows on the official account, the evidence suggests that these constituted an important source of net surplus outflow during the dynamic phase of agricultural growth prior to the First World War. Land taxes during the Meiji period formed the major share of government revenue, and were far greater than current and capital official transfers to agriculture. In the slow growth phase during the interwar period, on the other hand, the deceleration in the rate of increase in agricultural taxes, combined with increasing current and capital transfers by the government, implied a declining rate of surplus outflow from agriculture on official accounts, and, indeed, the 1930s witnessed a net inflow of official transfers into the agricultural sector.

Output and productivity growth in agriculture are not, of course, the only sources of income for the farm households. On the other hand, not all of the agricultural value added accrues to the farm households. The net factor income flows constituted an important source of surplus inflow into Japanese agriculture during the period under study. This is largely explained by the inflow of substantial wage labour income from the non-agricultural sector into the farm sector, indicating the importance of the rate and nature of growth of the non-agricultural sector for the intersectoral resource flow processes. The high rate of growth of the non-agricultural sector in Japan during the period under study, and particularly the high degree of labour absorption in this sector, created the conditions for the farm households to supplement their agricultural incomes with substantial wage labour income generated in other sectors of the economy.

A final feature of the intersectoral resource flows in Japan which needs to be highlighted here is the important role played by the voluntary savings channelled through the credit system as a source of financial outflow from agriculture. This signifies the fact that, in the context of growing farm household incomes, surplus outflow need not necessarily be forced through the exactions of the government or a parasitic absentee landlord class. In an economy with a relatively well-developed financial system, the outflow could take place through voluntary savings of the farm households, depending on the relative rates of return on investment in agriculture *vis-à-vis* the rest of the economy.

9

Intersectoral Resource Flows in China, 1952–1983

9.1 Introduction

The patterns and processes of intersectoral resource flows in China during the 1952–83 period are closely linked to the immense organizational changes which took place in Chinese agriculture over these years. Collectivization and centralized planning of production and distribution were among the key features of agricultural organization which greatly influenced the pattern of intersectoral resource flows until 1979. The collective organization of agriculture in this period played a central role in mobilizing idle resources, predominantly labour, in the rural economy of China for the purposes of capital construction. Some economists, however, have gone further in arguing that the central control over agricultural production through the new collective agrarian institutions, such as people's communes, helped to divert a substantial part of the agricultural surplus towards industrial accumulation. Though the question of the contribution of agricultural surplus to industrial accumulation will be examined in Chapter 10, in the present chapter we shall be addressing this issue from the point of view of the agricultural sector. One of the questions to be addressed here is the extent to which the empirical evidence supports the agricultural squeeze hypothesis. A key consideration in this regard is the degree to which the backwardness of Chinese agriculture constrained the government's attempts to extract a surplus from the sector while maintaining a basic standard of living for the agricultural population.

In the post-1979 reform period the Chinese economy entered a new phase in which the emphasis seemed to be shifting from primary accumulation based on collective mobilization of labour to the attainment of greater efficiency and improvement in productivity and product quality. This was the period of de-collectivization and the introduction of the farm household responsibility system, with increasing reliance on market relations as both a system of incentives and an organizing principle for resource allocation in agriculture. The implications of these changes for intersectoral resource flows during the formative years of 1979–83 will also be discussed in this chapter. The chapter is organized in the following way. In the next section we examine the empirical evidence on intersectoral resource flows from the real side. The section also discusses the issue of the terms of trade and the problems of the evaluation of intersectoral resource flows in real terms. An analysis of the determinants of intersectoral resource flows from the real side is provided in Section 9.3, where it is argued that population pressure on land and low productivity of labour severely constrained the opportunities for surplus extraction from agriculture during the Maoist period. The rapid growth of agricultural output and productivity in the post-1979 period increased the potential for surplus transfer from agriculture and could be said to have increased the autonomy of the state to enforce such a surplus transfer through its financial policies. The degree to which this potential was realized is discussed in Section 9.4, which examines the financial mechanisms of resource transfer. A summary and conclusions of the chapter are given in Section 9.5.

9.2 Resource Flows on the Real Side

A number of sources in Chinese and English have estimated intersectoral resource flows in China.[1] Ishikawa (1967*a*; 1967*b*; 1988), Lardy (1983), Ash (1991), and Sheng (1992) are amongst the English-language sources which discuss different aspects of intersectoral resource flows. We shall be using all these sources in discussing surplus flows from the financial side below. On the real side, however, the only systematic studies are those of Ishikawa and Sheng. Ishikawa (1967*b*; 1988) provides estimates for the period 1952–7 and for the years 1966 and 1980. Since Ishikawa's estimates are broadly similar to those of Sheng (1992), and as the latter source provides estimates for the entire 1952–83 period, our measures of resource flows on the real side are mainly based on these.[2] Sheng's estimates are based on the sectoral distinction between the agricultural and non-agricultural sectors rather than the institutional one between farm and non-farm households.

Intersectoral resource flows at current prices and at constant values are shown in Tables 9.1 and 9.2 respectively. At current prices, there seems to be a net inflow of resources into the agricultural sector of a rising magnitude throughout the period. Sales, or marketed surplus, which also include taxes in kind and compulsory sales to the government at administered prices, start marginally below purchases but grow much more slowly than the latter, thus giving rise to a widening gap in net resource flow during the reference period. Net surplus flow into the agricultural sector, starting from about 10 per cent of the total sales, increased rapidly to about 60 per cent of total sales in 1976–8. With the inception of the institutional reforms in the late 1970s, this trend was somewhat moderated, but net surplus inflow by the end of the period still stood at more than 45 per cent of the total agricultural sales.

[1] For a review of the Chinese sources, see Sheng (1992).

[2] The main difference between the estimates by Sheng (1992) and Ishikawa (1967*b* and 1988) is in their price series for purchases of the agricultural sector. Ishikawa uses the official price series for retail sales of industrial products in the rural areas. Sheng, on the other hand, distinguishes between prices of consumer goods and producer goods purchases, and since her consumer goods prices include food purchases of the agricultural households, her results show a lower level of improvement in agricultural terms of trade than Ishikawa's. This, however, does not change the results dramatically, and we could have used Ishikawa's estimates, with similar conclusions to those drawn in the text above (see e.g. Karshenas, 1989, pp. 59–66).

Table 9.1 Intersectoral resource flows, China, 1952–1983 (bn. yuan, current prices)

Year[a]	Sales (x)[b]	Purchases				Net resource flows (X − M)
		Total	Consumer goods	Producer goods	Government investment	
1952–4	16.62	18.27	15.71	1.94	0.62	−1.65
1955–7	20.16	23.36	19.06	3.26	1.03	−3.19
1958–60	24.30	34.48	22.34	8.33	3.81	−10.18
1961–3	21.81	30.00	21.82	6.33	1.85	−8.19
1964–6	30.80	36.92	26.15	8.19	2.57	−6.12
1967–9	33.57	41.14	29.82	9.43	1.90	−7.58
1970–2	36.46	53.13	34.95	15.04	3.14	−16.67
1973–5	45.36	66.35	41.95	20.33	4.07	−20.99
1976–8	50.74	81.04	49.76	26.42	4.86	−30.30
1979–81	83.68	123.95	85.29	33.92	4.64	−40.17
1982–3	117.40	170.26	126.19	40.59	3.48	−52.85

[a] Figures are average values over 3 years the last row is averaged for two years.
[b] Includes taxes in kind.

Source: Sheng (1992).

In considering the pattern of intersectoral resource flows at current prices, however, it is important to note that these flows were significantly affected by the administered prices on both the inflow and outflow sides. Since, for most of the period under study, the Chinese economy was organized as a centralized command economy, with major production and investment decisions in agriculture being dictated by the central authorities, the administered prices played a largely distributive role. The relative sizes of different resource flow categories at current prices, and their change over time, therefore, may have more to do with the behaviour of relative administered prices with little relation to relative productivities in different sectors. Thus, it is particularly important in the case of China to consider the current price flows in conjunction with resource flows in real terms, in order to separate the purely price movement effects from the quantity effects.

However, before considering the intersectoral resource flows in real terms (presented in Table 9.2), it is necessary to address the question of an appropriate base year or reference point price system. As can be seen from Table 9.3, the terms of trade moved rapidly in favour of agriculture during the period under study. The income gains through terms of trade improvement in agriculture, measured in 1952 base year prices, amounted to 62bn. yuans during the 1952–83 period as a whole, which was one and a a half times greater than the increase in the value of total agricultural sales (measured in 1952 prices) over the same period. Given such substantial terms of trade movements, the choice of the base year is clearly expected to affect the measure of real net intersectoral resource flows in a significant way. This question has been subject to a great deal of controversy in the literature. As prices during most of the study period were administered prices, it is

Table 9.2 Real intersectoral resource flows, China, 1952–1983 (bn. yuan, 1952 and 1964 prices)

Year	Sales (x)[a]	Purchases			Net balance (X − M)	
		Total	Consumer goods	Producer goods	1952 prices	1964 prices
1952–4	15.50	17.89	15.43	2.46	−2.39	2.29
1955–7	17.57	21.99	17.84	4.15	−4.42	0.64
1958–60	19.35	31.58	20.00	11.58	−12.23	−7.59
1961–3	12.98	24.60	17.62	6.98	−11.61	−9.33
1964–6	19.79	31.82	21.67	10.15	−12.03	−7.34
1967–9	20.97	36.69	25.01	11.68	−15.72	−11.34
1970–2	22.45	48.81	29.42	19.39	−26.36	−23.13
1973–5	27.16	61.80	35.35	26.45	−34.65	−31.09
1976–8	29.89	75.70	41.82	33.88	−45.81	−43.14
1979–81	40.37	108.66	67.54	41.12	−68.29	−66.61
1982–3	54.49	141.15	96.41	44.74	−86.65	−84.14

[a] In 1952 prices. All figures refer to 3-year average values; the last row is averaged for two years.
Source: Sheng (1992).

not clear which year, if any, could be regarded as an appropriate base year. As was mentioned in Chapter 2, various Chinese economists have attempted to adopt reference point prices based on the labour theory of value. As was pointed out there, however, given the lack of operation of markets and the severe restrictions on labour mobility imposed by the Chinese government during the study period, such attempts have been severely handicapped by the lack of commensurability of labour time to unit of output in different sectors. The schemes devised by the Chinese economists to overcome this have produced widely diverging results, which is clearly unsatisfactory.

Ishikawa (1967*b*) suggests 1952 as a base year, on the grounds that, compared to the other years in the study period, it comes closest to what may be considered as a 'normal' year. According to Ishikawa:

> The year 1952 is officially considered to be the year in which the national economy recovered the highest levels of pre-war economic activity. The first Five-Year Plan began in the following year, 1953, and market strain accompanying rapid investment increases appeared from the second half of that year, but 1952 was a year in which the supply and demand relations with respect to resources were relatively stable ... [and] moves towards socialization were not as yet conspicuous. For such reasons as these it is customary for students of China to consider the year 1952 the most normal year in the period of the People's Republic of China.

However, as he and others have pointed out, compared to the relative prices prevailing in the world market, the agricultural terms of trade in China in 1952 are believed to have been somewhat depressed. This supposition is based on historical

Table 9.3 Terms of trade of agriculture, China, 1952–1983 (1952 = 100)

Year[a]	Price indices		Net barter terms of trade (P_a/P_n)	Income terms of trade[b]
	Agricultural sales (P_a)	Agricultural purchases (P_n)		
1952–4	106.87	101.99	104.73	0.84
1955–7	114.68	106.19	107.98	1.64
1958–60	125.76	109.08	115.27	4.20
1961–3	168.76	121.93	138.51	6.74
1964–6	155.40	116.15	133.86	8.08
1967–9	160.09	112.14	142.76	10.99
1970–2	162.36	108.94	149.08	16.11
1973–5	167.00	107.35	155.56	22.09
1976–8	169.62	107.07	158.43	27.95
1979–81	206.80	113.70	181.91	48.90
1982–3	215.35	120.55	178.65	62.12

[a] Figures refer to 3-year averages.
[b] Bn. yuans at 1952 prices, calculated as $m_a\ (1 - P_m/P_x)$.
Sources: Tables 9.1 and 9.2, Ishikawa (1967*b*).

evidence which suggests that the agricultural terms of trade in 1952 were about 30 per cent lower than in 1936/7, when free market relations are said to have been operating both in the domestic economy and in relation to foreign trade. This seems to be confirmed by a comparison of the terms of trade in China and in the rest of the world (with 1936/7 as the base year), which is shown in Figure 9.1.[3] As the figure shows, assuming the two terms of trade to have been at par in 1936/7, the Chinese terms of trade show a decline of more than 39 per cent compared to their international counterpart in 1952. This gap closes rapidly, however, and, apart from the international commodity boom years of the early 1970s, the Chinese terms of trade seem to have stayed above the world series from the mid-1960s, developing a substantial gap by the end of the study period. As the world terms of trade show wide fluctuations during the period under study, taking world prices as the reference point prices would also be problematic. Given that the terms of trade in China in 1964 seem to come closest to the world level, and that this value is also close to the average of world terms of trade for the period as a whole (see Figure 9.1), it may be plausible to take this year as a base year. We have therefore measured the intersectoral resource flows at both 1952 and 1964 base year prices, and the results are reported in Table 9.2.

[3] In comparing these two series, it should be kept in mind that the world terms of trade refer to the index of agricultural commodity exports divided by exports of industrial products by industrialized countries, both measured on fob basis. Changes in transport costs are thus excluded, and the composition of industrial goods purchases by Chinese farmers may also be very different from those exported by industrialized countries. Keeping these two caveats in mind, the world terms of trade index may be regarded as an approximate index to the changes in world relative prices.

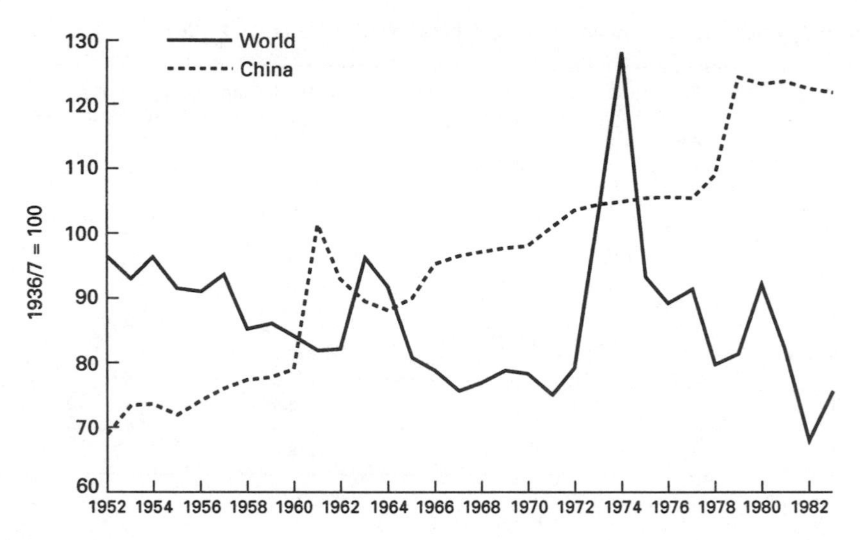

FIG. 9.1 Agricultural terms of trade, China and the world, 1952–1983

The estimates of real net intersectoral resource flows at 1952 base year prices seem to confirm the pattern observed at current prices. There was a net inflow of resources into the agricultural sector throughout the period, with a rapidly increasing trend. As a comparison between the figures in Tables 9.1 and 9.2 shows, the rate of increase in the real net surplus inflow and its level in terms of the total sales of the agricultural sector were indeed much higher than those suggested by the current value figures. This resulted from the rapid improvement in the terms of trade and was evidence of the substantial subsidies which the Chinese government granted to the agricultural sector by keeping the price of industrial goods coming into the sector stable in the face of increasing agricultural sales prices.[4] The net surplus inflow in real terms increased from about 15 per cent of total agricultural sales at the beginning of the period to 153 per cent in 1976–8 and 159 per cent in 1982–3 (see Table 9.2). These figures reflect sizeable resource transfers to the agricultural sector by any standards. A similar picture emerges if the measurements are made in 1964 base year prices (see Table 9.2). The trend towards increasing net surplus inflow to the agricultural sector is reproduced here, too, and in fact, with the exception of the early 1950s, when there is a small surplus outflow, the absolute value figures for the rest of the period remain similar to those obtained with 1952 as the base year.

[4] For example, the general index of retail prices of industrial products in rural areas for the years 1952, 1962, 1972, and 1982 was respectively 100, 115.4, 99.9, and 103.6. During the same years the general index of purchasing prices of farm and side-products was respectively 100, 164.6, 165.4, and 253.1 (State Statistical Bureau, 1987, pp. 78–9).

In addition to the net resource flow measures, Table 9.2 provides some interesting information on the pattern of resource flows which needs to be emphasized at this stage. An important point to be noted is that the sales of the agricultural sector throughout the period were not sufficient to finance even the consumer goods purchases from the other sectors of the economy. The gap was particularly notable during the post-1978 reform period, when one can witness a massive increase in the consumer goods purchases. During 1982–3, for example, the purchases of the agricultural sector for consumption purposes were almost 180 per cent more than its total sales. A second important feature of the pattern of resource flows is the changing role of the consumer and producer goods purchases of the agricultural sector in explaining the increase in total purchases during the pre- and post-1978 subperiods. While during the 1952–78 period about 55 per cent of the increase in the purchases of the agricultural sector was explained by the inflow of producers goods, during the 1978–83 period this share declined to only 17 per cent, with consumer goods purchases accounting for 83 per cent of the increase in the total inflows. These changing trends in the pattern of resource flows and the resulting net surplus flows were related to various factors on the real side – particularly those connected to productivity and output growth in the agricultural sector – as well as to mechanisms of financial resource flow transfer, which will be the subject of the next two sections of this chapter.

9.3 Determinants of Surplus Flow on the Real Side

Examining the trend of intersectoral resource flows from the real side involves the analysis of an interrelated set of factors affecting the sales and purchases of the agricultural sector. A central consideration in such an analysis is the delineation of factors which determine the inflow of resources into the agricultural sector – closely related to government policy in the case of China – and the productivity-enhancing role of these resources in agricultural production. A notable feature of intersectoral resource flows in China was that the ratio of agricultural marketed surplus, or total sales, to gross output was at a relatively low level, about 30 per cent, in the early 1950s. The low degree of commercialization of Chinese agriculture was indicative of the low productivity of labour and the lack of specialization in agricultural production. What is more noticeable is that the marketed surplus ratio remained more or less stable, with a very slow upward trend until the late 1970s. Total agricultural sales consisted of three broad components, namely, taxes in kind, compulsory sales to the government, and sales in the free market. Tax in kind was a major component of the marketed surplus during the 1950s, but its magnitude in both relative and absolute terms was continuously declining from the early 1960s onwards.[5] This decline, however, was largely compensated for by an

[5] Tax in kind declined from about 10% of total agricultural income in 1952 to about 4% in 1978 and 2.6% in 1983. The share of this tax in total government budgetary revenues declined from about 15% in 1952 to 2.5% in 1978 and 2.6% in 1982 (Sheng, 1992; Ash, 1991).

increase in compulsory sales to the government. The state grain purchases, for example, which stood at about 54 per cent of total grain procurement by the state in 1953, increased to more than 77 per cent of the total by 1979 (Lardy, 1983, p. 104). The third component was the sales to non-agricultural households in the free market, which consisted of the residual of the non-self-consumed agricultural output after subtracting total government procurements. The share of free market sales in total agricultural marketed surplus fluctuated around an average figure of 5 per cent throughout the 1952–78 period.[6] It was not before the post-1979 reform period that one could witness a discernible upward trend in the share of free market sales. Free market sales gradually increased from 5.5 per cent of total sales in 1978 to more than 10 per cent in 1983. By the early 1980s, however, the marketed surplus ratio was still less than 40 per cent of gross output.

Despite the fact that government procurements formed the major part of the agricultural marketed surplus during the study period, it would be misleading to conclude that the slow increase in total agricultural sales was due to deliberate government policy of maintaining low agricultural delivery quotas. The relative stability of the share of free market sales up to 1979 is indicative of the fact that the slow increase in agricultural marketed surplus had more to do with physical constraints imposed by stagnant agricultural productivity. An important factor was the mounting population pressure on land and the sluggish productivity of labour in agriculture. As is shown in Table 9.4, during the 1957–77 period, value added per agricultural labourer in fact registered a negative growth rate, −0.5 per cent per annum. Under these circumstances, the food consumption needs of agricultural households would increasingly encroach upon the marketed surplus and it would be difficult to increase the latter whether through state procurements or the free market.[7] Another factor responsible for the slow growth of the marketed surplus was the policy of regional self-sufficiency of food production pursued by the government from the mid-1960s. This would be expected, on the one hand, to hinder specialization and trade, and hence directly to reduce the marketed surplus ratio at any given level of agricultural productivity. On the other hand, by reducing agricultural productivity, the policy of regional food self-sufficiency would have exerted an indirect effect on retarding the growth

[6] Free market sales as a proportion of total sales stood at about 8% in 1952, declined to 6.5% in 1953, and remained at around 6% up to 1957. During the agricultural crisis years of 1958–60, it declined to an average of 2–3%. With the simultaneous revival of agricultural production and the decline in government procurements during 1961 and 1962, the rate increased to around 13%. However, as the government moved to increase its procurement, the rate declined rapidly and remained at around 4.5–5.5% for the rest of the period up to 1978.

[7] In this regard, it is important to point out that a large share of the government procurements (about 25%) throughout the period under study was resold to the agricultural sector in food deficit regions. If this is subtracted from the marketed surplus, the total sales ratio would be even less than the observed value of 30% mentioned in the text.

Table 9.4 Growth of output, land, labour, and productivity in Chinese agriculture, 1952–1979 (average annual growth rates)

Year	Value added (1)	Labour force (2)	Labour productivity (1 − 2)	Sown area (4)	Land/labour ratio (4 − 2)	Yield (1 − 4)
1952–7	3.7	2.3	1.4	2.2	−0.1	1.5
1957–78	1.5	2.0	−0.5	−0.2	−2.2	1.7
1978–83	6.3	2.0	4.3	−0.8	−2.8	7.1
1952–83	2.6	2.1	0.5	−0.6	−2.7	3.2

Sources: *Statistical Yearbook of China* (1987), and Aubert (1988).

of agricultural marketed surplus ratio.[8] This is part of a broader set of issues related to the efficiency of resource use in Chinese agriculture and the impact on land and labour productivity of the inflow of resources, to which we shall be turning now.

As we observed in the previous section, relative to the marketed surplus, commodity inflows to agriculture showed a more rapid rate of growth in real terms. This was due to a combination of reasons. On the one hand, as a result of terms of trade improvements, income and consumption in the agricultural sector, particularly of non-agricultural commodities, grew much faster than the real productivity growth in the sector would have warranted. As already discussed, net barter terms of trade almost doubled in favour of agriculture between 1952 and 1983. The income gains from terms of trade improvements were particularly significant in the post-1979 reform era. Such income gains, measured at 1952 base year prices, amounted to approximately 27bn. yuans in the twenty-six years between 1952 and 1978, and climbed rapidly by another 34bn. yuans in the five years between 1978 and 1983. Such substantial income gains would clearly allow a larger inflow of consumer goods purchases by the farm households from other sectors than could be financed by the internal resources of the agricultural sector itself. This, combined with the introduction of the household responsibility system in agricultural production during the 1979–83 period, seems to have been an important factor in the rapid acceleration of consumer goods inflow into the agricultural sector, as noted in the previous section. A further possible source of increase in demand for non-agricultural consumer goods in the farm sector, over and above that warranted by growth of real agricultural output, was the relatively large inflow of wage income from the non-agricultural activities of the farm households. As will be discussed in the next section, this was a major financing item in the intersectoral flow accounts.

The rapid growth of inflow of manufactured intermediate and capital goods

[8] On the impact of the policy of regional self-sufficiency on agricultural growth, see Lardy (1983, pp. 48–88).

inputs was another important cause of the relatively fast increase of imports into the agricultural sector. Two periods can be distinguished with regard to the intensity of new manufactured input use in agriculture. During the 1950s there was remarkable output growth, with moderate increases in new inputs. This was a period of recuperation, when agriculture was recovering from the devastation of the war while at the same time benefiting from institutional reforms and the expansion of cultivated land. With the post-'Great Leap Forward' crisis of agriculture in the early 1960s, food shortages posed a new urgency for the government's agricultural policy, leading to phenomenal increases in the inflow of mechanical and biochemical inputs to agriculture. The government often provided 30–50 per cent price subsidies on the use of new manufactured inputs in agriculture. Greater allocation of industrial investment to fertilizer and agricultural machinery plants also led to a rapid increase in the supply of such products, particularly in the latter part of the 1970s.[9] In the literature on Chinese agriculture it is often claimed that, in the sectoral allocation of investment by the government, the agricultural sector was relatively neglected. This argument is often supported by the claim that agriculture received only about 10 per cent of the centralized investment funds in the government budget, which is disproportionate to the share of agriculture in the GDP (see e.g. Lardy, 1983). There are, however, various flaws in this argument. First, it does not take into account the huge price subsidies on investment goods in agriculture. Taking these into account, the relative share of centralized investment funds allocated to agriculture in real terms was between 30 and 50 per cent higher than its current value. Secondly, it does not allow for agricultural investment financed by the internal funds of the communes and state farms. Thirdly, one should take into account the substantial budgetary allocations for investment in industrial activities for which the agricultural sector was the sole beneficiary, e.g. agricultural machinery and chemical fertilizers.[10] Finally, this argument neglects the great use that was made of the internal resources of the farm sector, and especially labour, in agricultural investment, particularly in construction and land infrastructure. To recognize the magnitude of investment effort in Chinese agriculture over this period, a comparison with the experience of other developing countries, based on the World Bank estimates, would be helpful. For example, the rate of use of chemical fertilizers (in nutrient units) in China was 90 kg/ha of arable land in 1978, while the same figure for other developing countries was only 25 kg. The trend growth rate of this figure for China over the 1965–78 period was 11 per

[9] For example, the total number of tractors in use increased from 1,300 in 1952 to 72,600 in 1965 and 745,000 in 1980. Hand tractors increased from 39,600 in 1965 to 1,874,000 in 1980. Accordingly, machine-ploughed area increased from 0.1m. ha in 1952 to 41m. ha in 1980. The use of nitrogen fertilizers also increased from 0.19m. tons in 1952 to 4.94m. tons in 1965 and 42m. tons in 1979.

[10] For example, during the 1963–80 period, about 10% of the overall investment in heavy industry was absorbed by investment in chemical fertilizer and agricultural machinery industries (Ash, 1991).

cent, compared to 9 per cent for other developing countries. Similarly, the amount of arable land per tractor in China in 1978 was 180 ha, with a trend rate of growth of 15 per cent, compared with 270 ha for other developing countries, with a trend rate of growth of only 7 per cent. In terms of investment in land infrastructure, also, China appears to compare favourably with other countries. For example, the share of irrigated land in total arable land was 45 per cent in China in 1978, with a trend rate of growth of over 3 per cent per annum over the 1960s and 1970s, compared to 17 per cent in other developing countries, with a trend rate of growth of only 2 per cent (see World Bank, 1983*a*, p. 25).

An important explanatory factor on the real side in the Chinese experience of intersectoral commodity flows seems to have been the lack of response in agricultural output and marketed surplus to the growing amount of resources allocated to the sector during the 1960s and the 1970s. As can be seen from Table 9.4, the rate of growth of agricultural value added during the 1957–77 period was well below that of agricultural labour force, leading to a negative productivity growth rate, −0.5 per annum. This was a time when there was a rapid growth of inflow of new inputs in the agricultural sector. The changing trends in output and productivity growth rates for the different periods shown in the table should be viewed in the context of the changing trends in land/labour ratio. The 1950s were a period of expansion of cultivable land. Over the 1952–7 period, total sown area increased by more than 16m. ha, or close to 2.2 per cent per annum. During the 1957–78 period, however, due to the encroachment of residential dwellings and industrial sites and the exhaustion of new agricultural lands, the total sown area declined by about 7m. ha, or −0.2 per cent per annum.[11] During the same period agricultural labour increased by 101m., or at an average annual rate of 2 per cent (see Table 9.4). As a result, the land/labour ratio, which remained more or less stable during the 1950s, declined by about 2.2 per cent per annum over the 1957–79 period.[12] In terms of growth of land productivity, therefore, the performance in the two subperiods was not very different – the annual rate of growth of output per unit of sown area was 1.5 per cent in the 1952–7 period and about 1.7 per cent in the 1957–78 period.

In comparison to the experience of other countries, and given that the performance was sustained over a long period of time, these appear to be quite respectable growth rates. The problem that remains, however, is that, despite the increased intensity of application of labour and other resources, the 1957–78 period still exhibited land productivity growth rates that were not significantly higher than the earlier period. This is tantamount to a decline in the overall productivity of the

[11] These figures are based on the data in the *Statistical Yearbook of China* (1987).

[12] This compares with a decline in land/labour ratio of about 0.7% per annum in India over a similar period. By the end of the 1970s, there were over 3 farm workers/ha in China, compared to 1 in India.

resources used in Chinese agriculture over the 1960s and the 1970s.[13] The decline in the productivity of resource use in Chinese agriculture over this period has been attributed to organizational problems such as over-centralization of decision-making processes, inefficiencies resulting from information asymmetries inherent in quantity planning, which was prevalent over this period, lack of incentives, lack of specialization resulting from the government's policy of regional food self-sufficiency, etc. (see Lardy 1983; World Bank, 1983; Perkins and Yusuf, 1984). The great surge in agricultural production after the reforms of the late 1970s and the early 1980s is indicative of the hindrance that such organizational problems had posed to the realization of the full productive potential of agriculture.[14] As far as the impact on the net intersectoral resource flows is concerned, however, as we observed above, the positive effects of the greater output and productivity growth rates in the post-1979 period was more than neutralized by the surge in consumer expenditure on non-agricultural products by the farm households. The introduction of the individual household responsibility system, combined with acceleration in output and productivity growth rates and large improvements in agricultural terms of trade, entailed a considerable increase in the income of the farm households. In the absence of an effective system of taxation of the farm households, and with a low degree of financial development, this increase seems increasingly to have found its way into consumer spending. This leads us to the discussion of the financial mechanisms of resource flow, which are the subject of the next section.

9.4 Financial Flows

The financial composition of intersectoral resources flows for the years 1956 and 1980 is shown in Table 9.5. The financial side of the intersectoral resource flows was very much shaped by the institutions of centralized planning during the study period in China. These institutions, which were aimed at rapid mobilization and centralized control over the investment of the economic surplus, were also significant for the financial flows into and out of the agricultural sector. In particular, the

[13] There are no comparable estimates of total factor productivity for the different subperiods in the period under study. The existing estimates, however, support the view that total factor productivity was particularly low during the 1957–79 period. Estimates by Tang (1984) indicate a negative rate of growth of total factor productivity for Chinese agriculture over the 1952–77 period of about 8%. According to one set of estimates by Perkins and Yusuf (1984), the contribution of total factor productivity to overall growth was also negative over the 1957–79 period. An alternative estimate by the same authors, based on a lower assumed marginal product of labour, suggests a 29% contribution of total factor productivity to the growth of output. Even this high estimate is well below those for Japan and Taiwan, discussed in the previous chapters. According to estimates by Fan (1991), about 42% of output growth during 1965–85 can be explained by total factor productivity, of which the major part was accounted for by institutional changes in the post-1979 period.

[14] According to the published data in the *Statistical Yearbook of China* (1987), an annual rate of growth of more than 10% was registered for gross agricultural output over the first half of the 1980s.

Table 9.5 Financing of net agricultural resource outflow, China, 1956, 1980, (bn. yuans, current prices)

	1956	1980
Net resource outflow[a] (R)	−5.09	−42.43
(*a*) Agriculture's sales (X)	19.22	84.22
(*b*) Agriculture's purchases (M)	−24.31	−126.65
Financing items		
1. Net outflow of factor income[b] ($F_a - Y_f$)	−7.45	−34.80
(*a*) Labour income[c]	−7.41	−34.79
3. Net outflow of current transfers ($T_{fg} - T_{gf}$)	1.87	−3.07
(*a*) Tax in kind	2.85	2.77
(*b*) Direct subsidies[d]	−0.98	−5.84
4. Net outflow of capital transfers	−2.95	−7.35
(*a*) Public capital[e] ($K_{fg}-K_{gf}$)	−1.21	−5.25
(*b*) Other capital[f] ($K_{fo}-K_{of}$)	1.74	−2.10
Financial institutions	−1.74	2.35
Non-agricultural enterprises	—	−4.45
5. Errors and omissions	3.44	2.79

[a] Refers to goods and non-factor services.
[b] Includes net interest outflow.
[c] Includes remittances of migrant workers as estimated by Sheng (1992, p. 134).
[d] Includes government finance for supporting the rural collective economy and rural relief funds.
[e] Government investment in agriculture.
[f] Includes capital transfers through financial institutions and by non-agricultural enterprises.

Source: Sheng (1992).

changing agrarian institutions had important implications for both the composition of the financial side of the intersectoral resource flows and the interpretation of the economic significance of these flows. In this respect, the period under study can be divided into two subperiods with two broad tendencies, namely, the 1952–77 period, or the Maoist period, which was one of growth and consolidation of collective organizations of agriculture, and the 1978–93 period, the period of decollectivization, which witnessed the gradual introduction of the household responsibility system, market incentives, and the dismantling of the collective organizations of agricultural production.[15]

A notable feature of the financial flows which prevailed throughout the study period was the significant contribution of factor income inflows, as is clearly shown

[15] This distinction is, of course, in relation to broad tendencies at work in institutional change. A deeper understanding of the economic significance of the different financial resource flow mechanisms entails a more detailed discussion of agricultural institutions in different phases during these two periods, which falls beyond the confines of the present study. For further analysis of organizational and institutional changes during the study period, see Lardy (1983), Perkins and Yusuf (1984), Stavis (1982), and Kojima (1988).

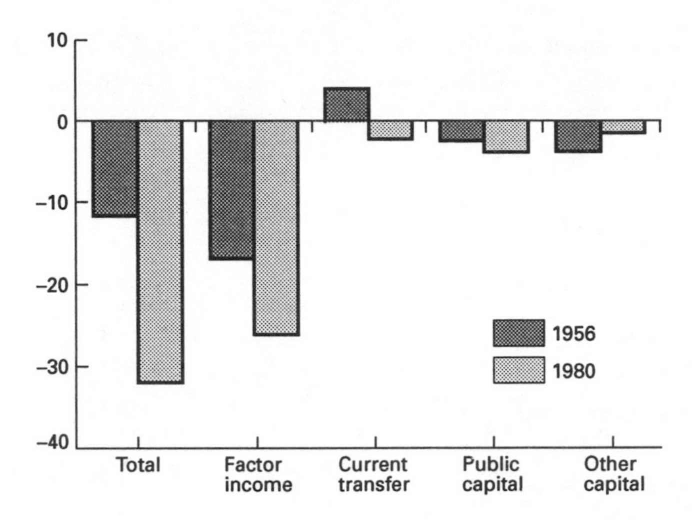

FIG. 9.2 Mechanisms of resource outflow from Chinese agriculture, 1956 and 1980

for 1956 and 1980 in Table 9.5 and Figure 9.2. In 1980 net factor inflow constituted more than 80 per cent of total net inflow of resources into agriculture, and in 1956 it was one and a half times higher than the total net inflow. In the absence of private landlordism, the net factor income inflow was almost entirely composed of labour income of farm households from non-agricultural activities.[16] The relatively high rates of labour income inflow were due to the growing importance of commune-managed enterprises in rural areas as a source of non-agricultural income. Compared to the case of India, discussed in Chapter 5, where factor incomes registered an outflow (based on Mundle's estimates) or a small inflow (based on our new estimates), this phenomenon highlights one of the organizational achievements of the Chinese economy in making productive use of the rural surplus labour. An important component of this income was generated in small-scale rural industries which catered for the input needs of the agricultural sector.[17] Rural surplus labour was also mobilized for large-scale infrastructural investment in agriculture during the Maoist period. During this period the government severely restricted labour mobility, and wage labour income was confined to the rural-based activities organized and managed by the government. In the post-Mao period, following decollectivization and the introduction of commercial norms, the inflow of wage

[16] As can be seen from Table 9.5, the other component of net factor income flows, namely, the net interest outflow, was negligible during the study period.

[17] It is estimated that, in 1977, for example, these industries produced over 40% of China's nitrogen fertilizer output, over 50% of phosphorus fertilizers, 64% of the cement, 33% of hydroelectricity, and almost all of the small and medium-sized machine tools. For a discussion of rural industrialization in China, see Wong (1982).

labour income to the farm sector accelerated.[18] This was connected to the rapid rates of growth of agricultural incomes and the resulting consumer boom in the first half of the 1980s, which encouraged a proliferation of non-agricultural commercial activities in the rural areas.[19] The continuation of this trend, however, in addition to the sustained growth of agricultural incomes, depends crucially on the survival of rural industries under competitive conditions and on the removal of the huge state subsidies which they enjoyed during the study period.[20] The removal of these subsidies, while maintaining high enough prices to ensure the survival of the small-scale rural industries, may turn out to be too costly from the point of view of the agricultural sector. On the other hand, the removal of subsidies and the introduction of competitive pricing could lead to the closure of some of the rural industries. This is one of the difficulties of restructuring the Chinese economy in the decollectivization period. The problem is how to move from a stage of development in which the emphasis has been on primary accumulation and productive employment of the surplus labour to one in which dynamic economies, quality improvements, and productivity gains are on the agenda, without risking a large-scale wastage of resources which may result from mass unemployment of labour. A gradualist approach, whereby market liberalization is synchronized with productivity improvement in the small-scale industries, seems necessary for a smooth transition.

Within the institutional setting of the Chinese economy, it would perhaps be more plausible to combine the analysis of the factor income flows with that of the current transfers. The reason for this is that various types of resource flows which in a market economy take the form of factor incomes – e.g. rents and profits – in the context of a centrally planned economy appear as public sector current transfers. There is a commonly held view, for example, that the state budget during the period under study played a central role in siphoning off a large part of the agricultural surplus through direct and indirect channels. Row 3 of Table 9.5 shows the net current transfers through the public sector accounts. As can be seen, the net current official transfers in 1956, though registering a positive outflow, remain relatively small compared to factor income flows. The outflow through

[18] Wage labour income inflow to the farm sector during the 1952–78 period increased by about 10bn. yuans, while during 1978–83 it increased by 12bn. yuans (both measured at 1957 retail prices of consumer goods in rural areas).

[19] It is estimated that rural industry gained 8.6m. new jobs in the 1980–9 period, with employment in the sector increasing by 36% from 24m. in 1980 to 32.6m. in 1989. Growth of employment in the rural services sector was even higher, rising from 11.1m. in 1980 to 52.5m. in 1989, or by 372% (Wiemer, 1992).

[20] In this context it should be noted, that despite the sharp increases in the purchase price of agricultural output during the 1978–83 period, the policy of subsidized agricultural inputs continued. As we have already noted in Section 9.2, this partly explained the substantial income gains resulting from the terms of trade improvements in favour of agriculture during the post-Mao period. Such subsidies were reflected in the losses made by the small-scale rural industries catering for the input needs of agriculture.

current official transfers declined continuously during the study period, and by 1980 it had already turned into an inflow. The reason for this, as we have already noted in the previous section, was the relative decline in the significance of tax in kind while direct subsidies to the agricultural sector were mounting. The importance of the official transfers as a source of surplus extraction from agriculture becomes even less pronounced if one also takes into account the capital transfers of the government to the agricultural sector. Inclusive of the capital account, net official transfers are reduced to only 0.66bn. yuan, or about 1.5 per cent, of agricultural value added in 1956, and in 1980 they record a net inflow of 8.32bn. yuans, or equivalent to 6.3 per cent, of agricultural value added.

It appears, therefore, that, contrary to what is commonly believed, official transfers as a source of surplus outflow from agriculture were either negligible or, as in later years during the study period, they constituted a positive inflow into the agricultural sector. It should be noted, however, that the official financial flow accounts in Table 9.5 refer only to visible flows at current prices.[21] An important source of resource transfer between the government and the agricultural sector during the study period was through the invisible account, or the terms of trade effect. For example, as noted above, large government subsidies to the agricultural sector on the input side were reflected in the terms of trade effect as a source of positive surplus inflow. On the other hand, it has been argued in the literature that the underpricing of compulsory sales of agricultural products to the government was an important source of taxation of agriculture (see e.g. Lardy, 1983; Kung, 1992). However, the net result of these invisible transfers, as discussed in Section 9.2, appears to indicate an inflow of resources to agriculture, especially during the 1970s and the 1980s. The net financial burden of the agricultural sector on central state accounts seems to have become a particularly serious problem during the post-1979 period of decollectivization. The government continued its policy of subsidizing consumer food prices as well as agricultural input prices, while agricultural procurement prices were raised substantially as an incentive measure in favour of the farm households. Such price subsidies increased from 4.9bn. yuans in 1978 to 25.4bn. yuans in 1983 (Kung, 1992, p. 143).[22] An important distinction between the experience of China in this period and that of countries such as Japan and Taiwan is that the Chinese government made no attempt to siphon off the rapid increase in the incomes of the farm households through direct taxation. Taxes in kind, which were the main form of direct taxation of agriculture, in fact

[21] A further qualification in considering the official transfers in the table is that they refer solely to the central government budget. They exclude, for example, the investment in agriculture at the commune level financed by retained earnings of agricultural institutions. Inclusion of these transfers would involve the addition of an outflow on factor income accounts and an equivalent inflow as investment by agricultural institutions in the agricultural sector. Such flows, however, have been netted out in the table, as they constitute an intrasectoral transfer of funds inside the agricultural sector.

[22] Of course, some of these subsidies benefited the urban consumers rather than the agricultural sector. A large part, however, accrued to the farm households through their food purchases in the retail consumer market in the rural areas and through their purchases of subsidized inputs.

declined in real terms from the very low level of about 2.79bn. yuans in 1978 to 2.59bn. yuans (measured in 1978 prices).

Net capital transfers registered an inflow into agriculture throughout the period. Though in absolute terms the magnitude of the capital inflows was increasing, relative to other financial flows and also in real terms it showed a pronounced decline in the post-1979 decollectivization period. This was the outcome of a number of processes related to the official and other capital flows. The public capital flows in Table 9.5 consist entirely of government investment in agriculture, and, in accordance with our practice in other chapters, financial flows through the banking system are shown as 'other capital'. Central government investment in agriculture during the Maoist period played an important role in the construction of large-scale infrastructural projects in the agricultural sector. The role of government investment was particularly important in water conservation and irrigation, which absorbed about two-thirds of the investment funds over the period as a whole. In the post-1979 reform period, however, as agriculture began to absorb an increasing share of government expenditure in the form of current subsidies, there was a precipitate decline in the proportion of government fixed investment allocated to agriculture. The share of agriculture in the overall capital construction investment by the government declined to 6 per cent in 1983 and to 3 per cent in 1985, after having been maintained at an average rate of about 10 per cent during the Maoist period.[23] Despite the rapid growth of farm household incomes after the introduction of the household responsibility system, this slow-down in publicly financed investment did not seem to have been compensated for by an increase in private investment in agriculture in the post-reform period (Ash, 1991).

The 'other capital' category in Table 9.5 consists of transfers through the financial institutions and capital inflows by non-agricultural enterprises. Credits made available through the financial institutions, namely, the People's Bank, the Agricultural Bank, and the rural credit co-operatives, were important sources of finance for agricultural investment during the study period.[24] Bank credits during the Maoist period were confined almost entirely to the collective agricultural

[23] There was in fact an absolute decline in government investment at current prices from 5.8bn. yuans in 1979 to 2.9bn. in 1981 and 3.5 in 1983. As pointed out in the previous section, central government investment was not the only source of publicly organized investment in agriculture. Other sources of public investment also witnessed a decline in the post-Mao period. For example, labour mobilization for agricultural capital construction fell from an average of 30–40 workdays per household per annum during the 1960s and the 1970s to 3 or 4 workdays in the 1980s. See Ash (1991) for more details.

[24] The significance of bank credit as a source of finance for agricultural investment underwent cyclical changes during the study period along with the changes in the mode of organization of agricultural production. During the First Plan, bank-loan financed investment in agriculture was equal to about 40% of investment financed by the central budget of the government. During the 1960s and up to 1977, which was the period of quantity planning, bank loans were drastically reduced in their role of financing agricultural investment. In the post-Mao period, however, this role increased substantially. The ratio of bank credit financing to central budget investment funds increased from about 17% in 1977 to 74.3% in 1983.

institutions and the state farms. During the post-1978 period, however, bank credits were increasingly put at the disposal of the farm households. Despite the sizeable increase in the value of credits granted to the farm households during the post-reform period, the net financial resource flows through the banking system indicated a relatively large outflow from 1980 onwards. This was due to the fast increase in the savings deposits of the farm households resulting from the high rates of growth of disposable income in the farm sector.[25] This source of outflow of funds was relatively small, however, compared to the other sources of resource inflow shown in Table 9.5, and, indeed, was easily overshadowed by the inflow of capital transfers by non-agricultural enterprises during the post-reform period. In comparing these results with those of other countries in our sample, particularly Japan, where capital flows through the banking system form an important source of resource outflow from agriculture (see Chapter 5), a number of considerations have to be taken into account. First, by the end of the 1970s, despite the achievements of the earlier two decades, Chinese agriculture was characterized by extremely high levels of population pressure on land and by low labour productivity and living standards. As a consequence, one would expect a large part of the increase in incomes during the post-reform period to have been absorbed by higher and more diversified consumer spending by the farm households.[26] Secondly, the post-1978 period witnessed important institutional reforms which essentially transferred the control over agricultural surplus from the collective agrarian institutions to the farm households. This, as we observed above, was reflected in the growing public sector deficit on the farm sector accounts and declining government investment in agriculture. Part of the increment in agricultural incomes in the post-reform period has therefore to be viewed as a transfer of investible funds which needed to be earmarked by farm households for working and fixed capital expenditure within the context of the new individual household responsibility system. As a consequence, net financial outflows arising from the increase of the farm household incomes and savings would be proportionately lower than would have been expected otherwise.[27] A third factor, also related to

[25] According to the data provided in the *Statistical Yearbook of China* (1987), per capita peasant household income increased from 133.6 yuan in 1978 to 355.3 yuans in 1984. Savings deposits of the farm households increased from 1.2bn. yuans in 1979 to 6bn. in 1980, and remained above 6bn. yuans to the end of the study period. Thus, the net financial outflows from the farm households through the banking system turned positive in 1980, and remained above 2bn. yuans for the post-reform period as a whole (Sheng, 1992, table 6.4).

[26] The purchased component of consumer spending by farm households, for example, increased from 33.5% of their income in 1978 to 44.1% in 1984 (*Statistical Yearbook of China*, 1987).

[27] For example, according to the statistics provided in the *Statistical Yearbook of China* (1987), the savings ratio of peasant farmers increased from about 15% of their income in 1978 to about 25 per cent in 1984. The absolute value of the net non-official capital outflows, however, seems to have been much lower than is indicated by these figures, largely because a growing share of farm household savings had to be diverted into financing the working capital needs of agricultural production with the gradual decollectivization of the agricultural sector over this period.

the institutional changes in the post-1978 period, was that this period witnessed substantial increases in labour mobility and rapid expansion of non-agricultural commercial activities in the countryside. As a result, one would expect a significant amount of direct investment by the farm households in non-agricultural activities, which would not be reflected in the financial flows shown in Table 9.5. The 2.79bn. yuan discrepancy between the financial accounts and the real accounts shown as errors and omissions for 1980 in the table, may partly be a reflection of this missing item.[28]

9.5 Conclusions

To sum up: the Chinese experience indicates a net resource inflow into agriculture throughout the three decades of centrally planned economic development. Contrary to the popular belief that central planning was a means of mobilization of agricultural resources for industrial accumulation, the finance contribution of agriculture to accumulation in non-agricultural sectors was negative. Though mobilization of surplus labour in agriculture appears to have made an important contribution to accumulation in the economy as a whole, the main beneficiary actually seems to have been the agricultural sector itself. Measured at current prices, net surplus inflow into agriculture remained positive throughout the 1952–83 period, and of a rapidly increasing magnitude. In real terms, with the possible exception of the early 1950s, we can observe a similar pattern of net resource inflow towards agriculture on an even more magnified scale. This was the result of the substantial terms of trade gains in agriculture, particularly from the 1970s onwards.

Given the high and increasing population pressure on land and the low productivity of labour, the Chinese government during the Maoist period was severely constrained in its attempts to extract surplus funds from agriculture. On the one hand, to ensure a basic living standard in the countryside, the government had to provided substantial food subsidies, often in interregional grain sales. On the other hand, to ensure an adequate growth of marketed surplus, it diverted an increasing amount of new inputs into agriculture at subsidized prices. The result was that agriculture became a net drain on other sectors in terms of surplus flows. Another notable aspect of Chinese experience is the substantial inflow of wage labour income into the agricultural sector. Compared with India, for example, this demonstrates one of the significant achievements of the Chinese economy in

[28] The discrepancy between the real and the financial accounts was not only confined to the post-reform period. As can be seen from Table 9.5, in 1956, for example, there was an even larger discrepancy between the two accounts. This mainly reflects the inadequacies of the data in the earlier period, particularly on the financial side (see Sheng, 1992; Ash, 1991). A further source of discrepancy is that the financial accounts constructed by Sheng do not provide estimates of increase in cash in circulation and hoarding in the agricultural sector.

mobilizing its surplus labour for collective investment in both agriculture and non-agricultural rural investment during the Maoist period.

The negative net finance contribution of agriculture to the Chinese economy during the Maoist period was indicative of the lack of commensurate response of agricultural production to the substantial flow of new inputs into the sector. This was reflected in measures of the contribution of total factor productivity to output growth which were well below those of other countries in our sample, such as Japan and Taiwan (see Chapters 6 and 8), and, according to some estimates, even negative. The institutional reforms of the post-1979 period seemed substantially to have improved the efficiency of resource use in agriculture, and at least in the short run, they induced rapid output and productivity growth rates. Though the fast rates of labour productivity growth in the agricultural sector in this period created the potential for extracting a surplus from the sector, the evidence suggests that in fact the negative financial burden of agriculture increased noticeably over these years. An important reason for this was that, paradoxically, the Chinese government chose to reduce its direct taxation on agriculture at a time when agricultural incomes witnessed substantial increases. This stands in sharp contrast to the experience of countries such as Japan and Taiwan, where direct taxes on farm household incomes were an important source of surplus outflow from agriculture.

10

Intersectoral Resource Flows, Economic Growth, Income Distribution, and Poverty

10.1 Introduction

In the last five chapters we have observed a variety of experiences in relation to the patterns of intersectoral resource flows in the process of development in the five country case studies considered in this book. In the case of China, India, and Iran, the evidence suggested a negative net finance contribution of agriculture, or, to put it in dynamic context, the agricultural sector became a growing financial burden on the rest of the economy, during the study periods in question. The opposite seemed to be the case in Taiwan and Japan – though the agricultural sector in Japan witnessed a positive and increasing net resource outflow up to the 1920s, during the 1920s and the 1930s this trend seemed to be slackening somewhat. A comparative overview of these different country experiences and their policy implications will be discussed in the concluding chapter of the book. However, before doing that, we shall try to extend our empirical investigation in order to examine the relation between the observed patterns of intersectoral resource flows and the variations in the experiences of growth and structural change in the countries under investigation. The main task of this chapter is to consider the different viewpoints on the role of agricultural surplus in the process of development discussed in Chapter 1 in the light of the empirical evidence provided by our country case studies. Though early debate on the role of agricultural surplus in economic development paid little attention to the relation between intersectoral resource flows, income distribution, and poverty, more recent literature has noted a close affinity between these issues – e.g. in the literature on structural adjustment and income distribution. In this chapter we shall also examine the relation between different patterns of intersectoral resource flows and income distribution and poverty.

The chapter is organized in the following way. In the next section we analyse the implications of the different patterns and processes of intersectoral resource transfer for industrial accumulation and economic growth. Particular attention will be paid to the savings-constrained models of industrial accumulation, in which the role of agriculture is seen as the provision of financial surplus for industrial accumulation in the early stages of development. Section 10.3 discusses the implications of intersectoral resource flows for income distribution and poverty. It is argued that, in order to understand the intricate links between intersectoral resource flows and income distribution, it is essential to go beyond the static full employment models in which sectoral outputs are largely perceived as substitutes, in the sense that greater output in one sector would imply contraction of other sectors. The main conclusions and some of the possible policy implications of the results are briefly discussed in Section 10.4.

10.2 Implications for Industrial Accumulation and Growth

A central theme in the debate on intersectoral resource transfer has been related to the savings contribution of agriculture to economic growth. As we noted in Chapter 1, some economists have argued that, in a predominantly agrarian economy in the early stages of industrialization, the agricultural sector must provide the initial funds for industrial investment. As we have observed in the individual country case studies, however, the net finance contribution of agriculture to industrial accumulation in most of the economies in the sample, and for much of the observation period, appears to have been negative. In this chapter we shall examine the extent to which this phenomenon has retarded the pace of industrial accumulation in the economies concerned, and consider the validity of the central role often given to savings mobilization in the study of intersectoral resource flows.

A remarkable fact about the economies in the sample is that they all achieved considerable success in increasing their rate of savings during their respective reference periods. As can be seen from Table 10.1, national savings as a percentage of each country's GNP increased by between 15 and more than 100 per cent during the respective observation periods.[1] Since the length of the reference period varies between countries, annual figures may provide a better index of relative performance in this regard. The average annual rates of increase of savings ratios for the different economies were 1.1 per cent for China, 3.3 per cent for India, 7.1 per cent for Iran, 0.3 per cent for Japan, and 1.1 per cent for Taiwan. Clearly, the performance of the different economies in raising their savings ratio bears little relation to their experience with regard to the direction and magnitude of agricultural surplus flows, discussed in the previous five chapters. It could even be said that countries with a negative finance contribution of agriculture, e.g. China, India, and Iran, have outperformed those with a positive contribution, i.e. Japan and Taiwan, in the task of overall savings mobilization.

It should be noted, however, that 'net finance contribution' of agriculture refers to the net resources made available by agriculture for non-agricultural investment, while the savings ratio is defined with reference to the economy as a whole. A more appropriate definition of the overall savings contribution of agriculture is the 'net

[1] As noted in the table, the savings ratios for China are shown as a proportion of net material product rather than GNP. This helps to explain the high values of savings ratio compared to other countries. The rate of increase in savings ratio, however, is on a par with other countries in the table.

Table 10.1 Rate of domestic savings and net agricultural surplus ratio, China, India, Iran, Japan, and Taiwan

	Savings ratio[a]	Net agricultural surplus ratio[b]
China[c]		
1952–7	24.3	4.7
1958–68	26.3	1.5
1969–78	31.2	−0.4
1979–83	30.7	−0.7
India		
1951–3	9.4	n.a.
1960–3	13.6	2.7
1968–70	15.8	−0.1
Iran		
1963–4	13.5	3.6
1971–2	21.6	6.5
1976–7	35.2	3.4
Japan		
1888–92	13.7	6.2
1908–12	14.3	7.3
1918–22	19.5	4.5
1933–7	15.7	2.3
Taiwan[c]		
1911–15	13.6	17.1
1921–5	23.8	17.5
1931–5	23.7	13.7
1951–5	18.7	16.8
1956–60	21.4	13.4

[a] % of GNP at market prices, unless otherwise specified.

[b] Agricultural value added minus the consumption of direct agricultural producers as a ratio of GDP. Valued at current prices for all countries except Taiwan, which is valued at 1935–7 prices.

[c] In the case of China, ratios are expressed as % of net material product, and, in the case of Taiwan, as % of net domestic product.

Sources: Lee (1971), Ohkawa and Shinohara (1979)), *Statistical Yearbook of China* (1987), Karshenas (1990*b*), Mundle and Ohkawa (1979), Mundle (1981), and Ohkawa, *et al.* (1982).

agricultural surplus' defined in Chapter 2. It refers to the income generated in the agricultural sector minus the consumption of direct agricultural producers. Net agricultural surplus as a ratio of GNP is shown alongside the savings ratios in Table 10.1. These figures are worth comparing in terms of both their magnitudes and their behaviour over time. First, the level of agricultural surplus ratio, as opposed to the overall savings ratio, was almost negligible for all the countries most of the time, with the exception of Taiwan and the early Meiji period in Japan. Secondly, and more importantly, the rate of increase in savings ratio bore little resemblance to the change in agricultural surplus ratio in all the countries in the sample. While the overall rate of saving was increasing rapidly, the agricultural surplus ratio was either declining sharply, or remained stable (in the case of

Taiwan). This suggests that the explanation for the behaviour of the overall rates of saving and investment in these economies should be sought in factors other than those directly related to agricultural surplus mobilization. What is certain with regard to the sources of supply of savings is that, with the exception of Taiwan and early Meiji Japan, agriculture seems to have made little contribution to the overall savings mobilization, and in some cases it played a negative role. Nevertheless, this did not prevent these economies from achieving high overall rates of saving which even outperformed Japan and Taiwan in this regard.

The above result may seem paradoxical in view of the fact that all the economies under study were predominantly agrarian economies in which the major part of the labour force was engaged in agriculture and related activities. Nevertheless, all the countries in the sample had already achieved a certain degree of industrialization in the initial stages of their respective reference period (see Chapter 4), and it was in fact the growth of the industrial sector which seems to have made a major contribution towards providing the funds for further accumulation. A comparison between Tables 10.1 and 10.2 shows a positive relation between industrial growth and the increase in the savings ratio.[2] One of the well-known stylized facts in economic development, at least since it was highlighted in Lewis's (1954) paper, is that savings ratio tends to grow *pari passu* with the increase in the share of the output of modern industry in national product. This may be explained on the basis of the organizational form of the modern industrial enterprise, in which retained profits form a major part of the accumulation funds; the higher the share of modern industry in national output, the higher the share of profits and thus, *ceteris paribus*, the savings ratio. This, however, neglects another significant source of growth of savings in modern industry which arises from the increase in unit profits normally associated with major productivity gains in industry. An important attribute of modern manufacturing is that fast rates of growth of output are normally accompanied by high rates of labour productivity growth. This is clearly shown in Table 10.2 for the five economies under study, in which labour productivity growth accounted for more than 50 per cent of the growth of industrial output in most of the periods. For any given level of industrial terms of trade and product wages in industry (i.e. industrial wages divided by the industrial price index), increases in labour productivity would be reflected in an equivalent increase in unit profits. This implies a higher savings ratio for any given share of industrial output in total national product. Increases in agricultural terms of trade and industrial product wages are mechanisms for redistributing productivity gains in industry amongst the industrial workers and agricultural producers. Comparing the data on productivity and real wage growth rates in the industrial sector in Table 10.2, it is evident that there were major gains in unit profits in all the

[2] Of course, a complete analysis of the behaviour of savings in the countries under study falls beyond the confines of the present study. The above discussion is focused on factors related to the sectoral growth performance of the different economies which are of direct relevance to the debate on intersectoral resource flows in savings generation.

countries over their respective reference periods, with the exception of Iran.[3] Though the adverse terms of trade movements against the industrial sector in Japan, India, and China retarded the growth of unit profits, in most cases they were not strong enough to wipe out the unit profit gains arising from the relatively large gap between the growth rates of productivity and real product wages.[4]

It appears, therefore, that the impressive increases in savings ratios in the countries under study were more the consequence of the growth of the industrial sector than of the contribution of agricultural surplus. This was partly accounted for by the increase in the share of output of modern industry in national income, and partly by the increase in unit profits in industry resulting from high productivity growth rates and relatively low real wage increases. In the case of Iran, the existence of oil revenues allowed this process to take place alongside high rates of real wage increases and high rates of consumption growth. This suggests that the emphasis on the savings contribution of agriculture, hitherto central to the debate on intersectoral resource transfer, may have been misplaced. The emphasis on the role of savings in this debate mainly reflects a static view of the intersectoral resource allocation process which is not uncommon in the development literature, i.e. the sectoral allocation of a given amount of investible funds to maximize current total national output. However, a dynamic point of view, i.e. considering the process of savings generation as an endogenous part of the process of intersectoral resource allocation, gives a totally different turn to the argument, as we have seen above. This is not to deny the crucial role that agriculture plays in the growth process, particularly at the early stages of industrialization, but rather points to the need for a shift of emphasis in assessing that role.

The role of agriculture becomes much more important when considered from the point of view of employment generation in the non-agricultural sector rather than of the overall savings and investment rates as such. A comparison between the experiences of China and India on the one hand, and Japan and Taiwan on the other, brings this point into clear relief. As we have seen in Chapter 4, the former

[3] In the case of Iran, the availability of substantial surpluses from the oil sector allowed relatively large increases in real wages to be combined with high rates of growth of the savings ratio.

[4] This holds for all the countries, with the exception of China. In the case of China, real product wages in industry grew at least 2 percentage points faster than labour productivity over the 1952–80 period as a whole. The slow growth of real wages reported in Table 10.2 was the result of the sluggish growth of agricultural marketed surplus and the increase in food prices. During the 1957–70 period, for example, real product wages grew by 4.1% per annum in Chinese manufacturing, while real wages (deflated by the consumer price index) declined by about 1% per annum (see Table 10.2). In a sense, this reflected the price paid by the Chinese industrial workers for the inflow of resources to agriculture through the terms of trade changes. The experience of India during the 1962–70 period was, to some extent, similar to this; while real product wages in manufacturing industry grew by an annual average rate of 2.1%, real industrial wages (deflated by the consumer price index) declined by 1% per annum (see Table 10.2). In this case, however, industrial productivity was growing at least 1.5 percentage points faster than real product wages, reflecting a continuous growth in unit profits over this period. In the case of Japan and Taiwan, also, real product wages growth was well below productivity growth in industry, reflecting rapid increases in unit profits over their respective reference periods.

Table 10.2 Growth of industrial output, productivity, and real wages, China, Japan, India, Iran, and Taiwan

	Average annual growth rates		
	Output	Labour productivity	Real wages[a]
China			
1952–7	18.0	12.7	4.7
1957–70	10.0		−0.9
1970–9	9.2	3.0	1.9
Japan			
1885–1900	4.9	1.8	0.6
1900–20	5.1	2.4	2.5
1920–37	6.8	5.2	1.3
India			
1953–62	7.5	3.6	1.8
1962–70	4.7	3.7	−1.0
Iran[b]			
1966–70	13.2	4.6	4.1
1970–7	17.3	10.2	11.0
Taiwan			
1911–25	8.5	1.7	0.4
1925–40	7.2	4.9	−1.4
1951–60	10.8	6.2	0.8
1961–70	14.7	8.0	3.5
1970–80	14.0	8.0	8.1

[a] Deflated by the consumer price index.
[b] In the case of Iran, the figures refer to the large-scale factory sector (with more than 50 employees).
Sources: Ohkawa and Shinohara (1979), Lee (1971), UN, *Yearbook of Industrial Statistics* (1965–87), Statistical Yearbook of China (1987), and World Bank (1983*a*).

countries suffered from growing population pressure on land and an inadequate rate of absorption of new additions to the labour force in the non-agricultural sector. The agricultural labour force in China increased by more than 120m., or 72 per cent, between 1952 and 1979, and in India it grew by more than 66m., or 65 per cent, between 1951 and 1971. The share of agricultural labour force in the total in India remained stable at around 70 per cent over this period, and in China it declined from 84 per cent to 74 per cent between 1952 and 1979. Fast rates of population growth were only partially responsible for the growing pressure on land; comparison with the experience of Taiwan, which had similar rates of population growth, shows that the low degree of labour absorption in the non-agricultural sector is perhaps a more important cause.[5] The low degree of labour absorption in the non-agricultural sector has traditionally been addressed in the development literature in terms of industrial policies of the government, techno-

[5] In Taiwan the share of the agricultural labour force in the total declined from about 70% in 1911 to 56% in 1960. It further declined from 43.5% in 1966 to only 21.5% in 1979 (the change in the system of

logical possibilities of factor substitution in industry, factor price distortions, etc. These are clearly relevant issues, but their relevance is predicated upon the possibility of procurement of an adequate 'wage fund' to sustain a larger non-agricultural labour force in the first place. As is shown in Table 10.2, even with apparently inadequate rates of labour absorption, real industrial wages in China and India followed a declining trend during the 1960s as a result of the inadequate growth of agricultural marketed surplus.[6] Additional increases in non-agricultural employment, which would have entailed further decline in real wages under these circumstances, may easily have proved socially and even physiologically intolerable.[7]

The next question, therefore, concerns the extent to which intersectoral resource transfers have affected the growth of output and marketed surplus of the agricultural sector in the economies under study. A comparison of the evidence on intersectoral resource transfers detailed in the previous country case study chapters with the rates of growth of agricultural output in the sample countries over different reference periods shows that, indeed, there is no clear-cut correlation between the direction and magnitude of resource transfer and growth of agriculture. For example, Taiwan showed a very rapid rate of growth in agricultural production despite a substantial net outflow of resources from the agricultural sector: agricultural output in Taiwan grew by 3.2 per cent per annum during the 1911–20 period, 4.2 per cent a year during the 1920–40 period, and 4.3 per cent a year during the 1950–60 period. China shows relatively fast rates of growth of agricultural output (4.5 per cent per annum) over the 1952–7 period, which was one of moderate resource inflow into agriculture, but the growth rate deteriorates to 2.2 per cent per annum during the 1957–70 period, which was one of heavy net resource inflow into agriculture. The growth rate recovered to 4.6 per cent per annum during the 1970–80 period. In India agricultural output grew by 2.9 per cent per annum during 1951–65, but it declined to 2.4 per cent during the 1965–70

industrial classification led to a 10% reduction in the share of agriculture compared to the data related to the earlier period).

[6] In this regard, it is important to note that the decline in real industrial wages in China and India in this period was not due to a slow growth of money wages or real product wages. Real product wages in China during the 1957–70 period grew by an average annual rate of 4.1%, well above the labour productivity growth in industry, and in India they increased by 2.1% per annum, close to labour productivity growth, over the 1962–70 period. The decline in real wages (deflated by the consumer price index) was a result of the increase in agricultural terms of trade, symptomatic of the inadequate rate of growth of agricultural marketed surplus. In the case of Taiwan, on the other hand, the slow growth of real wages was the result of a slow increase in real product wages, when agricultural terms of trade were declining in favour of industry.

[7] In terms of the inadequacy of the rate of labour absorption in modern industry, Iran falls into the same category as India and China, although the causes are totally different. In the case of Iran, the low degree of labour absorption in modern industry was due to industrial policies of the government rather than the inadequacy of the 'wage fund'. Easy availability of foreign exchange meant that urban food supplies could always be supplemented by imports in Iran. The result of the combination of these factors, i.e. abundant urban food supply and low labour absorption in modern industry, was that a

period, which was a period of relatively more intensive net inflow of resources into agriculture. In the case of Iran, with substantial inflow of resources into agriculture, output grew by relatively high rates of 4.3 per cent per annum during the 1963–77 period. In Japan, with an outflow of resources, agricultural output grew by 1.9 per cent per annum over the 1890–1920 period, but the slow-down in surplus outflow in the 1920s and the 1930s coincided with a deceleration in agricultural growth, with annual growth of agricultural output declining to 0.7 per cent.[8]

The lack of any simple correlation between resource flows and agricultural growth indicates the significance of the efficiency of resource use and the mode of utilization of resources allocated through intersectoral flows in the growth process. Indeed, on the basis of the empirical evidence examined in the previous section, one may support the hypothesis that the degree to which resources are productively utilized in the agricultural sector is itself an important determinant of both agricultural growth and the magnitude and direction of net resource flows. The experiences of Japan and Taiwan in this regard stand out here. They were the only two economies in our sample in which agriculture did not pose a constraint to the rate of growth of industrial employment[9] and, at the same time, made a net finance contribution to investment in other sectors. As argued in Chapters 6 and 8, the nature and pace of technological progress in agricultural production were crucial in determining the contribution of agriculture to economic growth in both economies. In both cases, the spread of land-augmenting technical progress based on new seed-fertilizer technology led to relatively fast rates of growth of labour productivity, with minimal demand on fixed capital investment. The effectiveness of new technology was bolstered by adequate and timely investment in irrigation and land infrastructure.[10] Basic investment in land infrastructure by the government was complemented to a considerable extent by private investment in small-scale labour-intensive projects with a high degree of utilization of the internal resources of the farm sector itself. The divisible nature of the seed-fertilizer technology made it accessible to small peasant holdings which formed the mainstay of farm organization in both economies.[11] The improvements in irrigation

substantial share of the increase in labour force in Iran was absorbed in the urban informal sector, with meagre productivity and low standard of living (see Radwan, 1975).

[8] The output growth rates are taken from the following sources: Lee (1971) for Taiwan; Mody, Mundle, and Raj (1985) for India; Karshenas (1990*b*) for Iran; Ohkawa and Shinohara (1979) for Japan; and World Bank (1983) for China.

[9] In the case of pre-war Taiwan, certainly, there was ample room for expanding non-agricultural employment and, at the same time, supporting even higher rates of growth of industrial real wages than were actually experienced. This is evident from the substantial foreign trade surpluses in agricultural products and large capital outflows from the colony over this period.

[10] In the case of Japan, the basic investment in irrigation had largely taken place in the pre-Meiji restoration period, and, accordingly, the burden of fixed investment in irrigation was less than for the other 4 countries (see Ishikawa, 1967*a*).

[11] Modern mechanical engineering technology, concerned with the application of mechanical power to field operations, was absent in both economies during their respective reference periods; it was after the 1950s that this type of technology became prominent there.

and drainage systems induced the development and use of more fertilizer-responsive, high-yield variety seeds, which in turn were effective in counteracting the rising cost of investment in land infrastructure, and thus maintaining investment incentives (Yamada and Hayami, 1979*a*). The growth of productivity of resource use in agriculture resulting from technological progress accounted for more than 50 per cent of the growth of agricultural output in both economies throughout their respective reference periods. Without such high rates of technical progress, it is inconceivable that a net resource outflow from agriculture could have been maintained over such a long period of time without severely damaging the growth of agricultural output.

Similar technological packages of new high-yield variety seeds and the use of biochemical inputs were also introduced in India and China over the 'green revolution' period. As noted in Chapters 5 and 9, substantial increases in the inflow of new producer goods inputs into the agricultural sector in these two countries from the mid-1960s were aimed at bringing about a similar type of land-augmenting technical progress. These attempts, remained largely unsuccessful, however, at least during the periods under study here; that is, agricultural growth was not commensurate with the cost of new inputs, and the outcome was a growing net resource inflow into agriculture while the food-supply constraint in the economy remained in force. If genuine technological progress is defined by the degree to which total factor productivity increases, then, in contrast to the cases of Japan and Taiwan, the experiences of China and India, at least at an aggregate level, do not appear to constitute genuine technological progress.[12] The differences in the degree of success in achieving genuine technological progress in these countries, derived from the differences in their agrarian institutions and government policies. There are a number of excellent comparative studies on this topic, a full treatment of which will lead us far afield (see e.g. Ishikawa, 1967*a*; Mody, Mundle, and Raj, 1985; Hayami and Ruttan, 1971). Though there are valuable lessons to be learnt from the experiences of Japan and Taiwan about organizational factors which were conducive to genuine technological progress in agriculture, a few cautionary notes are in order here about drawing too close a parallel which may neglect the differences between economies in their initial conditions. First, it should be noted that the productivity of labour in the agricultural sector in Japan and Taiwan at the beginning of their respective reference periods was well above that prevailing in the other countries. As a consequence, the capacity of the private farm sector to make the necessary investments to adopt the new technology was that much greater. Secondly, the rate of growth of population in China and India was at least twice as high as it was in Japan. Though Taiwan had a similar rate of population growth to China and India, the fact that it started from a base of much higher agricultural labour productivity places it in a different category, closer to Japanese initial

[12] This also applies to the case of Iran, though technical change in Iranian agriculture was of a different type, mainly directed towards extensive farming with a highly capital-intensive, labour-saving technology.

conditions than the other countries in the sample. The type of agrarian organization which stimulates genuine technological progress under these circumstances may be very different from the historical cases of Japan and Taiwan. In this connection, China, India, and Iran may have as much to learn from each other, and from their own past experiences, as from the historical experiences of Japan and Taiwan.

10.3 Implications for Income Distribution and Poverty

Income distribution and poverty in the developing countries are related to complex economic and socio-political processes which have given rise to a voluminous literature in development economics in recent years. In this section we are only concerned with those aspects of the problem which are most immediately connected to the patterns and processes of intersectoral resource transfers. Initially, it would be appropriate to examine a certain approach which has been gaining popularity in recent years in the study of intersectoral resource transfers and poverty. This approach consists of, first, estimating the composition of the poor and their concentration in different sectors, and then measuring the effect of intersectoral resource transfers on the basis of the way in which they affect the sectoral distribution of income, assuming that the *intra*sectoral income distribution remains intact (see e.g. Kanbur, 1987; Addison and Demery, 1987).

There are a number of serious problems with this approach. First, the implicit underlying theory is based on the assumption of full employment of labour and other resources, whereby output growth in one sector has to take place at the expense of output in other sectors. As we have already noted, complementarities between industry and agriculture form a very important dynamic aspect of the process of growth, especially in the type of economy that we have been studying here, with huge reserves of surplus labour. The impact of different patterns and processes of intersectoral resource transfer depends as much, if not more, on the productivity of resource use and on the generation of new income and employment as on the redistribution of a given level of aggregate income.

A further problem with this approach is that it implicitly assumes that the main source of income inequality lies in the intersectoral differences between average income levels. This is not what the existing empirical evidence suggests, however. Decomposition analyses for a number of countries show that income differences within sectors are much more important than income differences between sectors, with *intrasectoral* inequality typically accounting for 80–90 per cent of the total (see Fields, 1980). Our own decomposition analysis of the log-variance measure of income inequality for Iran also shows a similar pattern (see Table 10.3). As the table shows, more than 90 per cent of income inequality in Iran in 1977/8 seems to have been accounted for by intrasectoral inequalities; in other words, if total income were to be redistributed so that the average income in different sectors were equalized, there would be only a 9.5 per cent reduction in the log-variance measure of income inequality. This suggests that, if intersectoral resource transfers are to play a significant role in alleviating income inequality and poverty, they have, in one way or another, to affect the structure of production and income distribution and the productivity of resource use *within* each sector.

Given that significant changes in income distribution and poverty involve far-reaching changes in the structure of production and employment, models based on the past trends of a single country may not be capable of capturing the important aspects of the process. A comparative study of the experiences of different countries is perhaps the best way of examining the prospects and possi-

Table 10.3 Decomposition of household expenditure by occupation, sector of activity, and level of education of the head of household, Iran, 1977/8

	Mean expenditure[a]	Household share (%)	Gini coefficient	Log variance	Contribution to inequality (%)[b]
Occupation					
Employers	448	6.2	0.513	1.281	7.2
Self-employed	217	39.5	0.461	0.818	29.2
Public sector wage and salary earners	452	12.1	0.479	0.957	10.5
Private sector wage and salary earners	221	30.0	0.496	0.944	25.5
Others	239	12.3	0.594	1.851	20.6
Intraclass contribution to inequality	—	—	—	0.077	7.0
Sector of activity					
Primary	185	42.6	0.442	0.737	31.9
Industry	244	26.6	0.494	0.966	26.1
Services	385	18.1	0.480	0.992	18.3
Non-specified	417	12.7	0.504	1.099	14.2
Intra sectoral contribution to inequality	—	—	—	0.093	9.5
Education					
Illiterate	187	63.4	0.468	0.892	53.6
Primary school	298	15.4	0.469	0.905	13.2
High school	421	18.1	0.460	0.874	15.0
University education	1070	2.0	0.369	1.035	1.9
Non-specified	325	1.2	0.581	1.676	1.8
Intra group contribution to inequality	—	—	—	0.153	14.5
TOTAL	262	100.0	0.514	1.119	100.0

[a] Hundred rials per month.
[b] Based on decomposition of log variance.

Source: CSO (1977).

bilities of influencing income distribution and poverty through different patterns and processes of intersectoral resource transfer. Of the five countries in our sample, only three – namely, India, Iran, and Taiwan – offer adequate estimates of income distribution and poverty for such a comparative study over the reference periods considered here. However, the distinct experiences of these three economies, in terms of both patterns of resource transfer and income distribution, still allow valuable insights to be gained from such a comparison.

The Case of Taiwan

We shall begin with an overview of the trends in income distribution and poverty in Taiwan, which can be used as a benchmark against which the performance of the other countries could be assessed. There are a number of studies on income distribution and growth for post-war Taiwan (see e.g. Kuo, 1975; Galenson, 1979; Fei, Ranis and Kuo, 1978, 1979). Some summary statistics on income distribution and poverty for the period 1953–72 are shown in Table 10.4. As the table shows, income distribution in Taiwan, which was much less skewed than in many other developing countries, moved towards greater equality over this period. The combination of rapid growth in real incomes and improving distribution of income led to a remarkable decline in the poverty rate. By the mid-1970s, Taiwan had almost totally eradicated the worst cases of poverty.

The improvement in income distribution in Taiwan took place in the context of a rapidly growing economy undergoing rapid structural change. Real GDP per capita grew by an average annual rate of 4.1 per cent during the 1950–60 period, and by more than 6 per cent a year over the 1960–70 period. The composition of output and employment also changed dramatically over these years. The share of agricultural employment in the total declined from about 63 per cent in 1950 to less than 30 per cent by the mid-1970s, while the share of agricultural income in the total declined from about 27 per cent to 10 per cent over the same period. Exports as a percentage of GDP increased from 8 per cent in 1950 to 40 per cent in 1975, largely accounted for by the expansion of non-agricultural exports.[13] The decline in the share of employment in the farm sector was accompanied by a rapid increase in farm income per head, due to both labour productivity growth in agriculture and the fast expansion of non-agricultural incomes of the farm households.[14] By the mid-1970s more than 55 per cent of the income of farm households came from non-agricultural activities. The rapid rate of growth of demand for wage labour led to an unprecedented rise in real wages from the late 1960s.[15] Real average income of the wage-earning households increased by 123 per cent between

[13] The share of non-agricultural exports in total exports increased from 49% in 1962 to about 85% in 1975.

[14] Average farm household income in real terms increased by 5.5% per annum over the 1964–72 period (Kuo, 1975).

[15] Throughout this century, up to the 1960s, real wages had remained more or less stable, with even a declining trend in the post-war period. For example, real industrial wages, averaged over 1956–60, were

Table 10.4 Indicators of income distribution and poverty in Taiwan, 1964–1972

	Share of the household income		
	Bottom 40%	Top 20%	Gini coefficient
1953	11.3	61.4	0.558
1959	15.4	51.0	0.440
1964	20.3	41.1	0.328
1972	21.8	38.6	0.291
	% households below poverty line[a]		
	A	B	C
1964	35	55	80
1972	10	20	35

[a] The poverty lines A, B, and C refer to real annual incomes below T$20,000, T$30,000, and T$40,000 respectively.

Sources: Pang (1992) and Fields (1980).

1964 and 1972, compared to an increase of 87 per cent for all households over the same period.[16]

Clearly, the trend towards a more equal income distribution in Taiwan over this period was not the result of redistribution from the high- to the low-income sectors; if anything, the net intersectoral resource flow was away from the farm sector, i.e. the low-income sector. The changes in income distribution were rather the result of major structural transformations in the economy which led to a shift of labour from the low-income sector and opened up immense possibilities for supplementing the agricultural income of the farm households by income from non-agricultural activities. The agricultural sector played its part in this process by providing an adequate food supply for the rapidly increasing population[17] – in addition to a surplus which was exported. The fast rates of growth of labour productivity in agriculture meant that agriculture also made a net finance contribution to growth, i.e. savings transfer for investment, in the other sectors of the economy. A comparison with the experiences of India and Iran highlights the significance of these factors.

The Case of India

There is a relatively rich repository of data on personal distribution of income and expenditure in India since the 1950s, which has given rise to numerous studies on

only 8% more than those prevailing in 1911–15, and declined by more than 22% compared to their level in 1931–5 (Lee, 1971, p. 29).

[16] The data in the above paragraph are based on the following sources: Lee (1971), Kuo (1975), Fields (1980), and Fei *et al.* (1979).

[17] The average annual rate of growth of population in Taiwan was 3.9% during 1950–60 and 3.0% during 1960–73, constituting one of the highest population growth rates in the world.

Table 10.5 Indicators of income distribution and poverty in India, 1958–1969

	Gini coefficient of household expenditure distribution				
	1958/9	1960/1	1963/4	1967/8	1968/9
Rural	0.340	0.321	0.297	0.293	0.310
Urban	0.348	0.350	0.360	0.345	0.350
	% population below the poverty line				
	1960/1		1964/5		1968/9
Rural[a]	38		45		54
Urban[b]	32		37		41

[a] Rural poverty line is defined as Rs 15 per capita per month at constant 1960/1 prices.
[b] Urban poverty line is defined as Rs 18 per capita per month at constant 1960/1 prices.
Source: Bardhan (1974).

income distribution, and particularly rural poverty, in the literature.[18] Table 10.5 shows a number of summary statistics on the scale of poverty and the distribution of household expenditure in rural and urban areas for the period 1958–69. As can be seen, there are indications of a weak trend towards greater rural sector equality and increasing urban inequality, leaving the overall pattern of distribution more or less intact over time. Other studies, based on the distribution of household income, also indicate little change in the pattern of inequality, with perhaps a slight trend towards a more equal distribution (see Fields, 1980). However, the indices of poverty shown in the table, based on Bardhan's (1974) estimates, demonstrate a clear trend towards growing poverty – both in the rural and urban areas.[19] The most recent estimates of poverty in rural areas by Ahluwalia (1985), based on a similar head-count measure and using the same poverty line of 15 rupees (1960/1 prices) as Bardhan, also show an increase in rural poverty during the 1960s, but improvements later in the 1970s.[20] Of course, the behaviour of such poverty measures over time crucially depends on the definition of the poverty threshold in the base year, and particularly on the price index used for deflating the current values of household expenditure over time.[21] Though controversy over the appro-

[18] For a recent collection of papers with further references to the earlier literature, see Mellor and Desai (1985).

[19] The head-count measure of poverty may not be very satisfactory in this respect, as it does not take into account the changing intensity of poverty within the poor. However, estimates based on Sen's index of poverty follow similar trends in India in this period (see Ahluwalia, 1985).

[20] Ahluwalia's estimates indicate a cyclical pattern in poverty from the mid-1950s to the late 1970s: the population in poverty declined from 54.1% in 1956/7 to 38.9% in 1960/1, rose to 56.5% in 1967/8, and then declined to 39.1% in 1977/8, with no clear overall trend.

[21] In fact, the reason why the head-count measure of rural poverty in Table 10.5 increases, while the distribution of consumer expenditure moves towards more equality, is the higher rates of increase of the relative price index for the consumption basket of the poor.

priate price index and measurement methodology has been rife in the literature, the number of the poor, and the increase in this number over time, remains staggering whatever the definition of poverty and measurement methodology. For example, according to Ahluwalia's head-count estimates of poverty, the number of rural poor increased from about 128m. in 1956 to 138m. in 1960 and to 205m. in 1970, representing an increase of more than 60 per cent over the 1956–70 period as a whole. Despite the declining trend in poverty indices estimated by Ahluwalia for the 1970s, by 1977 the absolute number of rural poor was still about 50 per cent higher than in the latter half of the 1950s.

These changes in income distribution and poverty were taking place in the context of a relatively sluggish economy and particularly slow rates of labour productivity growth in agriculture and structural change in employment. The rate of growth of gross output in agriculture declined from 3.64 per cent a year during the 1950s to 1.68 per cent a year, well below the population growth rate, during the 1960–70 period. Labour productivity in agriculture, as measured by net output per agricultural worker, remained at the same rate throughout the three decades of the 1950–80 period, with virtually a zero trend rate of growth. This in itself could be regarded as an achievement, in view of the rapidly increasing agricultural labour force and declining land/labour ratios discussed in the previous section. As we have already observed, the rate of growth of agricultural labour productivity sets a limit to the total number which can be employed in other sectors with given real wages. It is not surprising, therefore, that the share of agricultural labour force in the total remained virtually constant at about 70 per cent during the 1950–70 period.[22]

The sluggish pace of structural change in the Indian economy in this period is in sharp contrast to the case of Taiwan discussed above. It comes as no surprise, therefore, that the literature on income distribution and poverty in India is dominated by factors related to *redistribution* of income – such as agricultural terms of trade and food prices, distribution of assets, government taxation, agrarian institutions, and market power, etc. – rather than the generation of new income and its distribution through different patterns of employment and remuneration of labour. This is not to imply that these do not exert an important influence on income generation and structural change, but it is their redistributive impact which is emphasized in the literature on income distribution and poverty. For example, important contributions to the literature on rural poverty have been on the institutional factors which condition the impact of the green revolution on rural income distribution and poverty (see e.g. Griffin, 1974; Griffin and Ghose, 1979; Bardhan, 1985). Institutional factors which influence the rate of diffusion of the new technology and its effectiveness in increasing agricultural productivity have figured less prominently in this literature. It is plausible to argue, however, that the impact of the green revolution on agricultural productivity in general

[22] The statistics cited in the above paragraph are all based on Chakravarty (1987, Statistical Appendix, pp. 103–27).

plays an equal, if not more important, role in the alleviation of poverty than the institutional factors which condition the appropriation of its benefits.[23] An important factor which has to be noted in this regard is that, in a country with the structural characteristics of the Indian economy, the effect of redistributive measures aimed at poverty alleviation may be severely constrained by the rate of growth of labour productivity in the agricultural sector. Any attempt at such a redistribution, in the face of the food-supply constraint posed by limited growth of agricultural marketed surplus, would be at least partially neutralized by price increases resulting from the excess demand for food. On the other hand, the rapid growth of agricultural labour productivity, even without further redistributive measures, may itself lead to reduction of poverty by enhancing the possibility of employment and income generation in other sectors of the economy, in addition to its direct income effects in agriculture and its favourable price effects. This is not to discount the importance of institutional changes and other policy measures specifically aimed at income redistribution, but rather to emphasize the need for even more radical institutional changes and other policies aimed at increasing the pace of technological progress and labour productivity growth in agriculture.

Under the conditions prevailing in Indian agriculture, i.e. limited land, fast population growth, and diminishing returns to labour, land-augmenting technical progress as characterized by the high-yield technology of the green revolution is essential in increasing agricultural labour productivity and alleviating poverty. The government, by targeting its irrigation and land infrastructure investment which is essential for the introduction of the new technology, and other policies such as taxation, asset redistribution, and education of the poor, can mitigate the undesirable distributional consequences of the new technology (see Hirashima, 1985, for a discussion of such policies in the context of pre-war Japan). Technological progress, of course, is not the only determinant of productivity growth in agriculture; the rate of absorption of labour by the non-agricultural sector is equally important. As we have seen in the previous section, however, the rate of absorption of labour by the non-agricultural sector in India during the period under observation was itself constrained by the slow increase in agricultural marketed surplus. One aspect of the policy problem under these circumstances, therefore, is to devise a set of measures which could at least help to break the vicious circle by increasing the rate of labour absorption by the non-agricultural

[23] The methodology used in the above literature to analyse the different influences on rural poverty is to regress a measure of poverty on income per head of rural population, food prices, and a time trend which is meant to capture the influence of other independent (institutional) factors on poverty (for more, and sometimes conflicting, results see Saith, 1981; Ahluwalia, 1985). The coefficient of output per head is sometimes used to test the 'trickle-down' effect of the green revolution. This is not entirely satisfactory, however, as productivity growth in agriculture influences the other independent variables in the equation, e.g. food prices. The coefficient of the time trend variable is sometimes used to test the impact of the green revolution once the output and price influences are removed. An adequate test of the overall impact of the green revolution, however, has to compare the actual trends with what would have happened in the absence of the green revolution (see Dantwala, 1985).

sector in the face of the food-supply constraint, without exerting undue pressure on the standard of living of the industrial workforce. This could set off a process of sustained structural change, whereby higher labour productivity in agriculture paves the way for a faster rate of technological progress, which in turn creates the possibility of a more rapid rate of labour absorption in non-agriculture, and so on. Increasing the rate of growth of industrial exports may be a possibility in this respect in the case of India. This could help to relax the assumption of a strict foreign exchange constraint which has been implicit in our argument so far, and contribute to the expansion of the industrial 'wage fund' by increasing the possibility of imports of agricultural products in short supply. Of course, it would be unrealistic to expect the same type of export performance in India as was observed in Taiwan from the mid-1960s – the extreme difference in size between the two economies implies that anything remotely approaching the performance of Taiwan by India would be hard to accommodate within the world economy. However, the evidence on the export performance of the Indian economy in the recent past suggests that there is ample room for stepping up the rate of growth of industrial exports as far as absorption by the world economy is concerned.[24] It should be noted that even marginal increases in industrial exports could help to increase the capacity for labour absorption in non-agriculture to a considerable extent, as part of the industrial 'wage fund' is expected to be supplied by the growth of labour productivity in the agricultural sector itself.

Of course, fast rates of structural change and growth are not necessarily a guarantee of rapid rates of labour absorption in industry and more equal income distribution. In particular, if the process of growth is led by industrial exports or non-agricultural primary product exports, and not accompanied by appropriate agricultural and industrial policies, it may encourage instances of extreme duality and increasing inequality. There are important lessons to be learnt from Iran's experience in this respect.

The Case of Iran

The development of the Iranian economy during the 1963–77 period was characterized by a combination of rapid rates of growth and structural change with increasing income inequalities. Table 10.6 shows a number of summary statistics on income distribution and poverty for the 1959–77 period in rural and urban areas. These statistics confirm the findings of a number of other studies which indicate a higher degree of inequality relative to international norms, as well as a significant trend towards worsening distribution over the period, in both the urban

[24] For example, the share of India in world exports, excluding minerals and fuels, declined from 1.15% in 1960 to 1.0% in 1965, 0.71% in 1970, and 0.69% in 1976. Its share in total exports of developing economies, excluding the major oil exporters, declined from 7.04% in 1960 to 5.57% in 1970 and 4.93% in 1976. For the source of these data and further elaboration of industrial export performance of India over this period, see Ahluwalia (1985).

Table 10.6 Indicators of income distribution and poverty in Iran, 1959–1977

	Share of household consumption expenditure		Gini Coefficient[a]
	Top 20%	Bottom 40%	
Urban areas			
Central Bank surveys			
1959	52.1	13.8	0.4681
1969	55.8	12.9	0.4797
1975	57.7	9.1	0.5383
CSO surveys			
1972	46.7	15.9	0.4128
1974	54.8	12.3	0.5152
1977	57.1	11.5	0.5210
Rural areas			
1963	42.4	19.4	0.3440
1970	46.6	17.1	0.4140
1976	52.2	13.5	0.4743

	Percentage of population below poverty line[b]		
	A	B	C
1972			
Urban areas	2	30	64
Rural areas	4	17	29

[a] The Gini coefficients and decile shares have been calculated by fitting cumulative distribution curves to the group data given in the official published data. Pareto distribution curves were fitted to the lowest and the highest open intervals and third-degree polynomials to the rest of the intervals. Methods of fitting the curves as in Kakwani (1980).

[b] The poverty lines defined by A, B, and C refer respectively to calorie intake of less than 75%, less than 90%, and less than 99% of requirements.

Sources: Karshenas (1990*b*) for inequality measures; Katouzian (1981) for poverty measures.

and the rural areas.[25] The head-count indices of poverty indicate that in 1972, though the worst cases of poverty (calorie intake below 75 per cent of normal requirements) were relatively rare, a sizeable proportion of the population, especially in the urban areas, fell below the poverty line (i.e. 99 per cent of calorie requirements). Unfortunately, estimates of poverty do not exist for other years, but, given the relatively fast rates of real income and consumption growth in the economy over this period, it is unlikely that the scale of poverty increased.[26]

These trends in income distribution were accompanied by fast rates of growth

[25] See e.g. Oshima (1973), Mehran (1975), Pesaran (1976), Pesaran and Ghahvary (1978).

[26] For example, the average household consumption of the bottom 40% of the household decile group increased at an annual rate of 1.65% in real terms during the 1959–71 period, and at a rate of 7.05% a year during the 1972–7 period. The growing inequalities in this period were not due to the slow income growth of the low-income group, but rather they were a consequence of remarkably high rates of increase of income of the top decile groups. For instance, the average household consumption of the top 20% increased by an annual rate of 4% during the 1959–71 period, and by 18.94% a year over the 1972–7 period.

and structural change, even more impressive than in post-war Taiwan. Real per capita GNP grew by more than 6 per cent a year over the 1963–70 period, and by more than 12 per cent a year during the 1970–7 period.[27] The share of agricultural output in non-oil GNP declined from about 30 per cent in 1963 to less than 15 per cent in 1977, while the share of non-agricultural employment increased from about 45 per cent in 1963 to about 68 per cent in 1977. There was in fact a decline in the agricultural labour force of about 13 per cent in absolute terms in this period, despite the rapid rate of growth of population of about 3 per cent per annum; this was not achieved by any of the other four economies in the sample in their respective reference periods, not even by Taiwan in the post-1950s period, when the agricultural labour force remained more or less stable. The trend of worsening distribution of income in the face of the rapid structural transformation and apparent shift of labour out of the low productivity sector, namely, agriculture, may seem paradoxical – particularly when contrasted with the experience of Taiwan. The two economies, however, exhibited totally different patterns of technological change, which was an important determinant of their divergent income distribution patterns.

In contrast to the agricultural policies in pre-war Japan and Taiwan, which emphasized land-augmenting technological progress, the main emphasis in Iran was on extensive farming with highly capital-intensive methods, which in some instances even involved the displacement of peasant farmers. The result is clearly reflected in the changing patterns of distribution of land holdings over this period, as shown in Table 10.7. The table brings into clear relief the rapidly worsening distribution of land holdings over the 1960–74 period. As can be seen, of the approximately 5m. ha of new land brought under cultivation between 1960 and 1974, about 84 per cent was concentrated in the top 20 per cent of the largest holdings, with the bottom 60 per cent of holdings accounting for only 4.1 per cent of the increment, and the bottom 40 per cent actually losing land in the process. Government policy of encouraging mechanization in agriculture also reduced the employment generation effect of agricultural growth. As a consequence, while land under cultivation increased by more than 25 per cent during the 1966–76 period, employment in the agricultural sector declined by 11.5 per cent, from 3.38m. in 1966 to 2.99m. in 1976. Even with such a rapid increase in the land/labour ratio, by the mid-1970s landless labourers and poor peasant farmers with casual seasonal employment, extremely low productivity, and low standards of living constituted more than 60 per cent of rural households (see Karshenas, 1990*b*, ch. 6).

The sluggish demand for labour in the agricultural sector, combined with a booming urban economy and the availability of ample supplies of food, led to a

[27] This was not solely due to the fast rate of increase in oil income; non-oil GDP per head also grew by over 6% a year during the 1959–72 period.

Table 10.7 Decile distribution of land by holding class, Iran, 1960–1974

Holding class	Size of holding (ha)[a]	Share of cultivated land		Share of the increase in cultivated land,
		1960	1974	1960–74
Lowest 40%	0–2	5.1	3.4	−0.4
Lowest 60%	0–5	14.0	10.9	4.1
Second 40%	2–10	28.7	25.0	16.7
Lowest 80%	0–10	33.8	28.4	16.3
Top 20%	10+	66.2	71.6	83.7

Note: The Gini coefficient of land holding distribution in 1960 was 0.6490.

[a] The 'size of holding' groups are approximations, since there is generally an overlap between decile and size classifications.

Source: Karshenas (1990*b*).

rapid increase in rural–urban migration.[28] A very small segment of the new entries into the urban labour force, however, was absorbed by the modern high-productivity industrial and services sectors. Like the agricultural policy, the industrial policy of the government in this period also encouraged a growing degree of mechanization in the modern corporate manufacturing sector, which showed an extremely low degree of labour absorption relative to the rapid investment and output growth in the sector (see Table 10.8).[29] More than 75 per cent of the increase in manufacturing employment over the 1963–77 period was absorbed in the small-scale informal manufacturing sector, mainly composed of small artisan workshops, with a stagnant technology and low productivity of labour.[30] This

[28] For example, during the 1966–76 period, urban population increased by an annual rate of 4.9%, while the rate of growth of rural population was 1.1% a year. According to the manpower survey of 1972, more than 50% of the employed labour force in urban areas was composed of migrant labour (though not necessarily all from rural areas).

[29] The corporate manufacturing sector, defined here as establishments with more than 50 employees, absorbed only 15% of the increment in the industrial labour force during the 1963–77 period, while it was granted more than 97% of the long-term subsidized loans by the specialized banks and was also the major beneficiary of government's protective policies. It is interesting to note that the medium-sized firms, i.e. those employing 10–50 employees, only contributed 10% of the growth of industrial employment (see Table 10.8). One would have expected that the combination of desirable features such as labour intensity of production processes with reasonable levels of labour productivity would have made these medium-sized firms ideal for generating productive and gainful employment in a labour surplus economy like Iran. Various features of the industrial policy of the government, however, militated against the growth of this sector. For example, they received only 2.2% of the long-term loans from the specialized banks, and there were few linkages between this sector and the dynamic corporate manufacturing sector. The large-scale enterprises in the corporate sector formed technological enclaves within the manufacturing sector, depending largely on imports for their raw material supplies and technology (for more details, see Karshenas, 1990*b*, ch. 7).

[30] The average size of the small-scale manufacturing shown in Table 10.8 was less than 3 employees per establishment, largely composed of family labour. Productivity of labour in this sector was

Table 10.8 Degree of mechanization and employment generation in the manufacturing sector in Iran, 1963–1977

	1963–7	1968–72	1973–7
Incremental capital/labour ratio (m. rials)[a]	815	2791	6473
Increase in total manufacturing employment (000s)[b]	297	380	785
% contribution			
Small-scale establishments[c]	72	77	79
Medium-scale establishments[d]	10	10	11
Corporate sector[e]	18	13	10

[a] Increase in gross fixed investment divided by the increase in total employment in large-scale manufacturing (i.e. more than 10 employees).
[b] Refers to total persons engaged in manufacturing in both rural and urban areas.
[c] Less than 10 persons engaged per establishment.
[d] Between 10 and 50 persons engaged per establishment.
[e] More than 50 persons engaged per establishment.

Source: Karshenas (1990*b*).

pattern of industrial growth led to an increasing duality in the economy, which was again in sharp contrast to the industrial growth in post-war Taiwan, where investment was concentrated in labour-intensive industries, with the major part of the industrial labour force actually located in the rural areas.[31]

To sum up: the three economies studied above followed distinct paths with regard to income distribution and growth, and in relation to the intersectoral resource transfers. There seems to be no simple, straightforward relation between the direction and magnitude of intersectoral resource flows and the changes in income distribution and poverty. In the case of Taiwan, substantial net resource outflow from the relatively low-income sector, namely, agriculture, was accompanied by fast trends towards greater equality in income distribution and rapid eradication of poverty. In Iran and India, despite the similarities in the direction of the net intersectoral resource flows, there were substantial differences in the patterns of structural change and the trends in income distribution and poverty. A significant aspect of the problem highlighted in the above country case studies is that the patterns and processes of intersectoral resource flows seem to be more important for income distribution and poverty in relation to the

extremely low, less than the wage rate in the large-scale sector, and the rate of growth of labour productivity over the period was almost zero (Karshenas, 1990*b*, ch. 7).

[31] A comparison with the growth of services employment in Taiwan is also illuminating. Lack of adequate productive employment in industry in Iran meant that a growing number of the urban labour force was absorbed in services, largely in the informal sector, such as petty trade, repair shops, etc. For example, the share of services in total employment in Iran increased by 100% over 1962–78, from 24% of total in 1962–3 to 34% in 1977–8. In the case of Taiwan, however, due to the rapid growth of productive employment in industry, the share of employment in the services sector remained more or less stable at around 28–29% during the entire 1953–75 period.

generation of new income and employment and the structural changes that they bring about than for the redistribution of a given income between different sectors. In particular, the type of technological changes which the external resources help to bring about in the agricultural sector, as well as the speed of technological progress as indicated by the increasing efficiency of resource use in agriculture, appear to be of the utmost importance in this relation. Of course, the changing patterns of income distribution and poverty have other complex determinants, some of which are directly linked to the mechanisms of intersectoral resource transfers in ways other than those we have emphasized above. These can be underlined if the processes of intersectoral resource transfers are considered from the financial side. For example, official current transfers, i.e. government taxes and subsidies, which, as seen in the case studies, have been important mechanisms of intersectoral resource flows, could also play an important redistributive role in the economy. Again, the distribution of productive assets in agriculture and, more generally, the structure of agrarian relations play an important role in resource flows through net factor transfers (e.g. rent paid to non-cultivating landlords), as well as having a significant influence on income distribution and poverty. Official transfers on the capital account also have important implications for intersectoral resource flows as well as income distribution and poverty (e.g. through the targeting of government investment, as mentioned above). Similarly, the terms of trade mechanism, while being an important channel of intersectoral resource transfer, has obvious implications for income distribution, and particularly poverty. It is important to note that these mechanisms of resource transfer and income redistribution could also have important implications for the generation of new income and employment and for the process of structural change in the economy as a whole. For example, land taxes which take into account the differential productivity of different types of agricultural land, apart from being a redistributive mechanism, can also play a considerable part in increasing efficiency of agricultural land use (see Kaldor, 1962).[32] Similar dynamic effects could result, in varying degrees, from the appropriate use of other mechanisms of resource transfer, e.g. government investment, price policy, asset redistribution, etc., as we have already discussed. However, in the absence of such dynamic effects, and in a slowly growing economy, particularly with sluggish productivity growth in agriculture, the room to manipulate these instruments in order to change the direction of intersectoral resource flows or the patterns of income distribution and poverty is bound to be extremely limited.

[32] In this regard, it is important to note that land taxes, an important government revenue in the Meiji period in Japan and in Taiwan, are conspicuous by their absence from Iran and India, and, more generally, from other present-day developing countries.

10.4 Concluding Remarks

In this chapter we have examined the interrelationships between intersectoral resource flows, growth and structural change, and income distribution and poverty. The empirical evidence suggested that the net finance contribution of agriculture, at least in the surplus labour economies that we have been studying, does not seem to matter significantly in the process of savings generation in the economy as a whole when considered in a dynamic setting and from a long-term perspective. The growth of the savings ratio in the countries under study seemed to have little to do with surplus transfer from agriculture, and in fact, if anything, the relation between agricultural surplus outflow and the savings ratio appeared to be negative. Even in economies in which agriculture made a negative net finance contribution, great success was achieved in raising the overall savings ratio as a result of the increasing share of industrial output in national income and the relatively fast rates of productivity growth in the manufacturing industry.

It appears, therefore, that, according to the development experience of the countries studied here, the emphasis on the savings contribution of agriculture, hitherto central to the debate on intersectoral resource transfer, may have been misplaced. The central role of savings in this debate mainly reflects a view of the intersectoral resource allocation in which the emphasis lies in the choice of optimal sectoral allocation of a *given* amount of investible funds with *given* rates of return in different sectors. The empirical evidence examined in this chapter, however, suggests that, from a long-term perspective, the efficiency of resource use within the sectors, as determined by organizational and technological capabilities of the different economies, plays a more crucial role in the process of growth and structural change. This shift of emphasis from allocation of a given amount of resources between sectors or activities to the more dynamic considerations of the efficiency of resource use, notably connected to the process of technological progress, renders the magnitude and direction of net intersectoral resource flows endogenous to the growth process, rather than a policy variable as in the savings-constrained models of industrial accumulation. In developing countries with limited cultivable land, growing population pressure, and diminishing returns in agriculture, the possibilities for achieving significant land-augmenting technical progress through the green revolution are of utmost importance. This no doubt requires a growing *gross* inflow of resources, in the form of new biological inputs, investment in irrigation and land infrastructure, investment in human capital, etc., to the agricultural sector. Whether the final outcome leads to a *net* inflow of resources to the agricultural sector or outflow from the sector depends on the efficiency of the utilization of these resources in agriculture and the achievement of genuine technological progress in the sector. Net intersectoral resource flows, central to the savings-dominated models of growth, cannot be regarded as a policy instrument as such, therefore, but rather as an intermediate indicator which helps in the assessment of the degree of efficiency of resource use in agriculture.

The key contribution of technological progress in agriculture to economic growth in the countries in the sample seemed to be its effect in raising the productivity of agricultural labour. This is a crucial link in the process of growth and structural change in large agrarian economies, in which sustained growth in the industrial 'wage fund' is essential for the maintenance of a growing non-agricultural labour force. As the empirical evidence reviewed in this chapter suggested, the degree of success in implementing land-augmenting technological progress in the agricultural sector of the labour surplus economies is important not only because of the contribution it makes to the removal of the wage constraint; it also plays a crucial role in the changing patterns of income distribution and poverty and in the structural transformation of the economy as a whole. The divisible nature of the new seed-fertilizer technology allows the spread of benefits of technological progress amongst the small peasant holdings which constitute the core of operational units in the agricultural sector of the developing countries. Of course, government policy directed towards the creation of an appropriate environment, such as the establishment of relevant agrarian institutions, clear definition of property rights, security of tenure, and, above all, provision of infrastructural investments and supporting services, is an important precondition for reaping the distributional benefits of the new technology. In fact, such policies are essential for the successful diffusion of the new technology in the first place.

As we observed in the case of Taiwan, the growth of labour productivity in agriculture may lead to further beneficial distributional effects by increasing the demand for the products of labour-intensive industries and, in particular, encouraging the growth of such industries in rural areas, which may create an additional source of labour income for the farm sector.[33] The growth of labour productivity in agriculture is also a precondition for the alleviation of the worst types of poverty in large agrarian economies which face the food-supply constraint. As the comparison of the cases of India, Iran, and Taiwan indicated, if such productivity growth is based on genuine technological progress which leads to cost reductions in agricultural production, it can be doubly beneficial to the eradication of poverty; on the one hand, by preventing food price increases, and, on the other, by generating new sources of employment for the poor.

These crucial dynamic aspects of the relation between income distribution and intersectoral resource flows seem to have been totally neglected in the recent analytical models on intersectoral resource flows and income distribution. Such models, which have become popular in recent years in the discussion of the impact of structural adjustment on income distribution and poverty, essentially focus on the impact of a redistribution of given resources between different sectors in a comparative-static-type analysis. The static nature of these models, which arises from their basic assumptions of full employment and static technology, abstracts

[33] The existing empirical evidence from other countries also suggests that the increments in the income of low-income peasant households are in fact largely spent on the products of such labour-intensive industries (see Mellor, 1989).

precisely from those aspects of the link between income distribution and intersectoral resource flows which, as we have observed, are of crucial significance in the long run. A notable aspect of the relation between intersectoral resource flows and income distribution which is highlighted in the case studies in this chapter is that the patterns and processes of intersectoral resource flows seem to be more important for income distribution and poverty in relation to the generation of new income and employment and the structural changes that they give rise to than for the redistribution of a given income between different sectors.

11

A Comparative Overview and Conclusions

11.1 Introduction

The role of agricultural surplus in the process of industrialization has been the subject of considerable debate in the development literature. As we noted when reviewing this literature in Chapter 1, the different theoretical standpoints and their policy prescriptions crucially depend on the assumptions made about the structure and functioning of the economies in question. The assessment of the different policy prescriptions on the direction and magnitude of the net intersectoral resource flows suggested in the literature thus becomes primarily an empirical question. The comparative study of the processes of intersectoral resource flow in the five countries considered in this book was conducted with a view to shedding light on this debate within the context of empirically given economic structures. Our findings, however, are not solely of interest to the old debate on the financing role of agriculture in the early stages of development. They also have significant relevance to the more recent policy debates on trade and industrial policy and structural adjustment in the developing countries, as well as being of more general interest with regard to the processes of structural change and growth. This chapter is concerned with some of these broader implications of the intersectoral resource flow patterns observed in the country case studies of the earlier chapters. We shall begin by providing a comparative overview of the pattern of net intersectoral resource flows in the sample countries.

11.2 A Comparative overview

The five countries studied can be broadly divided into two groups on the basis of their experience of agricultural surplus transfer. The first group, namely, the *surplus group*, includes Taiwan and Japan, where the finance contribution of agriculture to the rest of the economy was on the whole positive and agricultural surplus was growing over time.[1] China, India, and Iran fall in the second category, the *deficit group*, where the net finance or net product contribution of agriculture to the other sectors was negative, and over time the agricultural sector became a growing financial burden on the rest of the economy. This result may seem

[1] For the definition of 'net finance contribution of agriculture' and other notions of agricultural surplus, see Ch. 2.

paradoxical, given that in all the countries in the latter group strong industrialization policies of a highly protective type were pursued, and that such policies in the development literature are normally associated with a resource squeeze from agriculture. To explain this apparently paradoxical result we must first compare the working of the various mechanisms of intersectoral resource flow in the countries under review. Such a comparative study was carried out in detail in the country case studies; in this chapter we shall be mainly concerned with issues of a more general theoretical and policy interest.

In order to facilitate our comparative study, a schematic picture of the financial composition of intersectoral balances for the countries studied is provided in Table 11.1. The positive and negative signs in the table are only indicative of the dominant tendencies in the direction of surplus flow during the reference periods for each country; for a more elaborate analysis of the different mechanisms of surplus flow on an annual basis one has to refer to the individual country case study chapters. A positive sign is indicative of a net surplus outflow from agriculture and a negative sign indicates a net inflow of resources into the agricultural sector. The countries in the table are grouped into the surplus and deficit categories discussed above, with Taiwan and Japan showing a positive net finance contribution by agriculture, followed by the other three countries with negative contributions.

The Case of Taiwan

Taiwan, at the top of the table, exemplifies the agricultural surplus case where the agricultural sector made a positive and growing net finance contribution to the other sectors throughout the 1911–60 period. This was not because of the low or

Table 11.1 Mechanisms of net resource outflow from agriculture

Country	Terms of trade (TT)	Factor income (F_a-Y_f)	Current government transfers ($T_{fg}-T_{gf}$)	Capital transfers		Total ($R=X-M$)
				Government ($K_{fg}-K_{gf}$)	Other[a] ($K_{fo}-K_{of}$)	
Surplus countries						
Taiwan[b]	+	-	+	-	+	+
Japan	-	-	+	-	+	+
Deficit countries						
China	-	-	+	-	-	-
India	-	-	+	-	-	-
Iran	-	-	-	-	-	-

[a] Other capital in this table refers to private capital transfers including credits granted by government banks to agriculture. For definition and detailed description of the column title headings, see Ch. 2.

[b] In the case of Taiwan, factor income flows and terms of trade in this table refer mainly to the post-colonial period.

declining level of the gross inflow of resources into the agricultural sector; on the contrary, Taiwanese agriculture witnessed a significant and growing inflow of resources into the sector. What gave rise to the sizeable and increasing value of net agricultural surplus outflow was the larger and faster-growing marketed surplus or sales of the agricultural sector relative to its purchases. This was made possible, on the one hand, by the relatively efficient resource use and fast productivity growth rates in agriculture, and, on the other hand, by the existence of effective mechanisms of surplus flow on the financial side. Effective government action both in transforming the technological basis of agricultural production and in siphoning off a large part of the fruits of productivity growth in agriculture through various financial mechanisms was, amongst other things, instrumental in the surplus transfer.

The need of the Japanese colonial administration to procure a sizeable exportable agricultural surplus from Taiwan played an important part in technological transformation of Taiwanese agriculture prior to the Second World War. Up to the 1920s, agricultural policy was mainly directed towards institutional reform, education, and infrastructure building. The growing population pressure on land and the rising demand for rice and sugar in the Japanese market necessitated large increases in land productivity and the transformation of the technological basis of Taiwanese agriculture. Heavy investment by the government in irrigation and land improvement during the 1920s facilitated the move toward intensive farming based on increasing use of fertilizer and new varieties of seed. This investment, together with other infrastructural investments during the early decades of this century, formed the foundation for the rapid spread of the new crop technologies and productivity growth in agriculture, which carried on into the post-colonial period as well. A distinct feature of Taiwanese agriculture, shared only by Japan amongst the countries in our sample, was that the major part of output growth during the period under study was accounted for by the growth of productivity of resources deployed in the sector. This created the potential for agricultural surplus outflow on a substantial and increasing scale.

The financial channels through which this transfer took place are depicted schematically in Table 11.1. Official capital transfers, which mainly consisted of government investment in agriculture, were the only channels of resource flow which consistently registered a net surplus inflow throughout the study period.[2] This was, however, dwarfed by the sizeable outflows through other channels. During the colonial period the main mechanisms of surplus outflow were through rents and government taxation. In the post-war period, rents lost their significance as a major mechanism of resource transfer, particularly in the aftermath of the land reform of the early 1950s, as government taxation and adverse terms of trade movements took over as the predominant sources of resource transfer. In fact, in

[2] It should be noted that the net inflow of factor income depicted in Table 11.1 only refers to the post-war period. For more details, see Ch. 6.

the post-colonial period one can observe a net inflow of resources through factor income flows, as a result of a substantial wage income inflow and with rents losing their significance. However, this was easily overshadowed by other sources of surplus outflow. The terms of trade, which up to the mid-1930s moved in favour of agriculture, showed a substantial deterioration against agriculture in the post-war period. The income squeeze on the farm sector through the terms of trade movement in the post-war period were as large as 60 per cent of the increase in incomes resulting from real value added growth in the sector during the same period. This rapid deterioration in agricultural terms of trade, as noted in Chapter 6, resulted from the government's pricing policies in its compulsory rice collection and fertilizer sales.

The terms of trade effect was supplemented to a major extent by government direct taxation, which was another important source of surplus outflow. Taxation was in fact so high that, despite the significance of government investment in agriculture, the net consolidated government accounts showed a sizeable and growing net outflow of resources from agriculture. In other words, the outflow of resources through taxation was far greater than the inflow through capital investment and subsidies, and the difference grew over time. What stands out in the case of Taiwan in this respect is that the tax burden on agriculture seems to have been much higher than on the other sectors of the economy.

Another source of surplus outflow from agriculture, particularly significant in the post-war period, was through private capital transfers, which primarily took the form of outflow of voluntary savings by the farm sector through the credit system. This signifies the fact that agricultural surplus transfer need not necessarily take the form of compulsory government or landlord extraction; under the conditions of rising productivity and incomes in the agricultural sector, surplus transfer can take place through voluntary savings, if the appropriate financial institutions are in place.

A notable feature of the experience of Taiwan was that the substantial squeeze on agriculture through taxation, rents, and other mechanisms did not seem to retard the rates of technological advance and output and productivity growth in the sector. In the post-colonial period in particular, large increases in government taxation and substantial adverse terms of trade movements were combined with high rates of growth in output and productivity growth in agriculture. As noted in Chapter 6, this was evidence of the fact that, in a technologically dynamic agriculture, productivity growth can help to maintain relative profitability, and hence the inducement to invest in agriculture, despite the adverse terms of trade movement.

The Case of Japan

The financial contribution of agriculture to the government budget during the Meiji period in Japan has become a classic example in the development literature of the financing role of agriculture in the early stages of development. Our empirical investigation of the case of Japan in Chapter 8 does support the conventional wisdom, and indeed more; in addition to its net contributions to the government's budget, the overall net finance contribution of agriculture to other sectors during the study period also seems to have been positive and considerable. We distinguished between two phases in the behaviour of net agricultural surplus during the study period. The first phase (1888–1917), exhibited large and expanding net agricultural surplus outflow, while in the second phase (1917–37) one could observe a deceleration in the rate of surplus outflow, even, according to some estimates, turning to an inflow. This behaviour in the rate of change of agricultural surplus flow reflects two distinct phases in the development of Japanese agriculture, with different productivity and output performances.

The first phase, from the Meiji restoration to the end of the First World War, was a period of relatively high productivity and output growth in the agricultural sector. As noted in Chapter 8, agricultural output outstripped population growth in this period, and, in addition to providing food and raw materials for the rapidly growing non-agricultural sectors, agriculture also produced an export surplus which made a significant contribution to industrial development. More than three-quarters of output growth in this period was accounted for by productivity growth rather than the increase in inputs, which indicated the fast rates of technological innovation and the high and improving degree of efficiency resource use in Japanese agriculture. Productivity growth in the agricultural sector during this phase was based on a continuous incremental improvement in technological practices within the traditional agrarian institutions inherited from the Tokugawa period. Technological innovations in the form of land improvement and extension of irrigation, as well as the introduction of new seeds, fertilizers, and better methods of cultivation, took place within smallholder family units with, on average, one hectare of land per household. The government and the rural-based landlord class played an important role in these technological improvements both by providing the finance for bulky investments and by promoting technological innovations and their diffusion. The rapid rates of productivity growth during this early phase created an increasing potential for agricultural surplus outflow.

The interwar period, which forms the second phase of development of Japanese agriculture in the period under study, was one of agricultural stagnation and low productivity and output growth. During this period, productivity growth played a minor role in the growth of agricultural output, and the limited growth which did take place was mainly accounted for by the increase in inputs. The sources of technological improvement within the traditional institutions of Japanese agriculture seem to have been exhausted by the end of the first decade of the twentieth century, and returns to investment in the sector within the traditional agrarian institutions were therefore diminishing. This meant that the rate of increase of resources directed towards agriculture also slowed down compared with the earlier period, and agricultural output stagnated. The slow-down in the growth of agricultural output and productivity meant that the role of agriculture as a potential source of surplus became increasingly limited. This was indeed the underlying cause of the reversal of trends of net agricultural surplus flow during the interwar period.

The composition of surplus flow on the financial side, as depicted schematically in Table 11.1, shows the different mechanisms through which this took place. The difference in the rates of productivity and output growth between the two sub-periods itself had important implications for the movements in the terms of trade and other mechanisms of financial flow. During the earlier, dynamic phase of agricultural growth, the terms of trade improved only slowly in favour of agriculture and agricultural protection remained moderate. Indeed, during the first fifteen years of the twentieth century, the agricultural terms of trade in Japan relative to world prices declined substantially. This was, however, rapidly reversed during the interwar period, as the terms of trade in Japan improved considerably in favour of agriculture relative to the agricultural terms of trade at the international level. With the slow-down in the rate of productivity growth in the interwar period, the government had to intervene to protect agricultural incomes, and the degree of protection of the agricultural sector increased substantially.

The current official transfers formed an important source of net surplus outflow during the dynamic phase of agricultural growth prior to the First World War. Land taxes during the Meiji era formed the major share of government revenue, and far exceeded the current and capital official transfers to agriculture. However, during the slow growth phase in the interwar period, the deceleration in the rate of increase in agricultural taxes, combined with increasing capital and current transfers by the government, led to a declining trend in the rate of surplus outflow from agriculture, which turned into an inflow towards the end of the period. As in the case of Taiwan, what stands out in Japan with regard to the pattern of official transfers, is the much higher burden of direct taxation on agriculture compared with non-agriculture. As discussed in Chapter 8, despite the narrowing of the gap between the relative tax burdens over the period under study, by the end of the 1930s the burden on agriculture was still twice as high as on the non-agricultural sector.

An important source of income transfer into the agricultural sector throughout the study period in Japan was the net factor income flows. These were due to the combination of low factor income outflow through rents, and large inflows in the form of labour income from the non-agricultural activities of farm household members. The former was due to the peculiarity of the agrarian relations in Japan, with its rural-based landlord class whose rental income did not constitute a factor income outflow from agriculture. The second factor, the wage income of agricultural households from non-agricultural activities, constituted a major source of income transfer into Japanese agriculture throughout the years under study. This exemplified the importance of the rate and nature of the growth of the non-agricultural sector for intersectoral resource flows. The high rates of growth of the non-agricultural sector in Japan during the study period, and especially the high degree of labour absorption within this sector, created the conditions for the farm households to supplement their agricultural incomes with substantial wage income generated in other sectors of the economy.

A final important feature of the Japanese experience of intersectoral resource flows was that, as in the case of post-colonial Taiwan, voluntary savings channelled through the credit system constituted an important source of financial outflow from the agricultural sector. While such net private capital outflow in the context of an economy at stationary state implies disinvestment and stagnation in the agricultural sector, in the context of a technologically dynamic economy undergoing rapid structural change it signifies the complementarity of sectors in the process of development and is normally accompanied by high rates of investment and productivity growth in agriculture.

The Case of China

Chinese agriculture underwent immense organizational changes during the period under study. During the 1952–79 period, collectivization and central planning of production and distribution were among the key features of agricultural organization with an important bearing on the processes of intersectoral resource flow. During the post-1979 reform era, the dismantling of collective organizations of agricultural production and the gradual introduction of the household responsibility system and market incentives meant that the pattern of intersectoral resource flows underwent considerable change. Despite this changing pattern of channels of resource flow, the net finance contribution of agriculture remained negative throughout the study period, with the gap widening over time. As noted in Chapter 9, with the possible exception of the early 1950s, this result held both at current value and in real terms, at different base year or reference point prices, including world prices. It appears, therefore, that contrary to the common belief about central planning being a means of mobilization of agricultural surplus for industrial accumulation, the finance contribution of agriculture to accumulation in non-agricultural sectors in China was negative.

Throughout the Maoist period, particularly during the 1960s and the first half of the 1970s, massive amounts of resources in the form of intermediate and capital goods were diverted to Chinese agriculture. A significant share of industrial investment in this period was allocated to heavy industries producing capital and intermediate inputs for the agricultural sector, with their products sold to agriculture at highly subsidized prices. During this period China demonstrated some of the fastest rates of diffusion amongst the developing countries of new agricultural inputs such as new seeds, fertilizers, pesticides, and mechanical implements. By the end of the 1970s China had achieved rates of use of modern inputs per hectare of cultivated land well above the average in the rest of the developing countries. However, output response to the growing amount of resources allocated to the agricultural sector during the 1960s and the 1970s was relatively low. Yields did not increase commensurate with the growth of the new inputs, and indeed, labour productivity exhibited negative growth over these two decades. Given these trends in agricultural productivity, the government was severely constrained, and became increasingly so over time, in extracting a surplus from agriculture. With the declining productivity of labour in agriculture, the government had to provide substantial consumer subsidies, such as food subsidies in interregional grain sales, in order to ensure a basic living standard in the countryside. Furthermore, given the inefficiency of resource use in Chinese agriculture, the government diverted an increasing amount of new inputs into agriculture at subsidized prices to ensure an adequate growth of output and marketed surplus. Consequently, agriculture became a drain on other sectors in terms of net resource flows.

The inefficiency of resource use in Chinese agriculture was partly due to over-centralized and inflexible agrarian organizations and their associated incentive and information problems. The rapid growth of output in the post-1979 reform period, which was largely based on the more efficient use of existing resources rather than increased flow of inputs, created the potential for a reversal in the growing trend of negative agricultural surplus flow. This, however, did not materialize as the higher income and productivity growth in agriculture in this period were more than neutralized by the surge in consumer expenditure by the farm households. As pointed out in Chapter 9, the share of consumer goods in the growth of total agricultural purchases during the 1978–83 period jumped to 83 per cent, from a value of slightly under 45 per cent during the 1952–78 period. The factors underlying this changing composition of intersectoral resource flow may be clarified further by investigating the financial mechanisms of resource flow as depicted in Table 11.1.

An important source of real income gain in the agricultural sector throughout the study period was the terms of trade effect. As discussed in Chapter 9, the income gains through the terms of trade effect during the 1952–83 period amounted to more that one and a half times the increase in the value of agricultural marketed surplus over the same period. Such substantial income gains were the result of the indirect subsidies provided by the government, through its policy of keeping the price of manufactured inputs supplied to agriculture constant while agricultural purchase prices increased over time. Such substantial income gains allowed a much larger inflow of producer and consumer goods purchased by agriculture from other sectors than could be financed by the internal resources of the agricultural sector itself. A major part of the gains in income terms of trade was concentrated in the post-1979 reform era, which played an important role in the rapid acceleration of consumer goods inflow into the agricultural sector.

There were, of course, other factors responsible for the growth of demand for non-agricultural consumer goods in the farm sector, over and above that warranted by growth of real agricultural output. The most important of these was the relatively large net factor income inflows into the farm sector, which predominantly took the form of wage income from the non-agricultural activities of the farm households. During the Maoist period, the relatively high rates of labour income inflow were due to the growing importance of commune-managed enterprises in rural areas as a source of non-agricultural income. In the post-1979 reform era, following decollectivization and the introduction of commercial norms, the inflow of wage labour income to the farm sector accelerated. While the continuation of government subsidies to rural enterprises kept the old sources of factor income flow to the farm households intact, new sources of income were added due to the rapid rates of growth of agricultural incomes and the resulting consumer boom in the first half of the 1980s which encouraged a proliferation of non-agricultural commercial activities in the rural areas. Factor income flows, when added to the income gains through the terms of trade effect, constituted a considerable inflow of external income to the farm sector, particularly in the post-1979 period.

In the context of a centrally planned economy, various items which in a market economy take the form of factor incomes – e.g. rents and profits – can appear as public sector current transfers. There is indeed an established view which maintains that the state budget played an important role in siphoning off a large part of the agricultural surplus during the period under study. As shown in Chapter 9, however, net current official transfers, though registering a positive outflow, remained relatively small compared with factor income flows, and declined continuously during the study period, even turning into an inflow by 1980. It is significant that during the post-1979 period, when agricultural incomes were rising rapidly due to the processes mentioned above, government direct taxation of agriculture indeed registered a negative growth in real terms. This was in sharp contrast to Japan and Taiwan, where government taxation of agriculture in the early stages of their development played an important role in generating a net positive finance contribution from agriculture. The lack of development of financial markets and the uncertainties regarding property rights and future developments in the economy helped to translate the income rises in the farm sector in the post-reform era into a surge in consumer spending. In this way, the widening financial resource gap of the agricultural sector continued even in the post-reform period, as already noted. While in the earlier period the growing financial burden of agriculture was due to the inefficiency of resource use and declining agricultural productivity, in the post-reform period the continuation of this trend was more due to a lack of appropriate institutions and mechanisms of resource transfer.

The Case of India

India is another agricultural deficit country in our sample in which the net finance contribution of agriculture to other sectors during its reference period was negative. Agricultural resource balance in the 1950s showed a relatively large deficit which followed a declining trend up to the mid-1960s. During this period there was an increasing trend in the value of marketed surplus as a proportion of agricultural value added, while purchases of the agricultural sector as a proportion of agricultural value added remained stable, thus creating a narrowing gap in the agricultural resource balance. From the mid-1960s, however, there was a sharp reversal in these trends, leading to a rapidly growing net inflow of resources into the farm sector. The growing financial burden of agriculture during the 1960s was closely connected to the 'new agricultural strategy' adopted by the Indian government.

The new strategy was introduced at a time when traditional agriculture seemed to be faltering because of the virtual exhaustion of new cultivable land frontiers and growing population pressure on land. Central to the new agricultural strategy was the seed-fertilizer technology of the green revolution which involved the introduction of a combination of new inputs into the agricultural sector, with obvious direct implications for the flow of resources. This, however, was not in itself a cause of increasingly negative net finance contribution by agriculture. As seen in the case of countries such as Japan and Taiwan, periods of growing positive net finance contribution by agriculture coincided with intensified gross inflow of resources into the agricultural sector. What led, in the case of India, to different behaviour between the net and the gross flow of resources was that agricultural supply response was not commensurate with the increased flow of new inputs into the sector, as evidenced by the rapid rise in the ratio of producer goods purchases to agricultural value added during the 1960s.

The inflow of new producer goods into the farm sector, however, was only one element of the purchases of agriculture from outside the sector. An even more important element was the increased flow of consumer goods into the farm sector connected to the income gains in the sector through the terms of trade effect. An important aspect of the new agricultural strategy was the considerable improvement in the agricultural terms of trade over the 1964–70 period. This was intended as an incentive device and also as a means of income transfer to make the adoption of the new technology affordable to the farm households. The income gains through the improvement of agricultural terms of trade during the 1964–70 period were more than twice those resulting from the normal growth of agricultural output over the same period. Such income gains through the relative price changes, as discussed in Chapter 5, exerted a double squeeze on the agricultural surplus: on the one hand they increased the demand for consumer goods from outside the farm sector, and on the other they reduced the net sales of farm products to other sectors. This, of course, presupposes that such income gains were not siphoned off from the farm sector through various mechanisms, such as factor income, and official and private current and capital transfers.

As far as the net factor income flows are concerned, the most plausible estimates seem to suggest a net inflow into the farm sector, although of a relatively much smaller magnitude than the other countries in our sample. Estimates of capital

transfers, both official and private, also suggest a net inflow into agriculture – the former being mainly government investment in agriculture and the latter taking the form of subsidized loans by the banking system to the farm sector. While in Japan and Taiwan, at least in later stages of their development, such bank credits to the farm sector were substantially overshadowed by the savings deposits of farm households, this does not seem to have occurred in the case of India. Given the relatively lower per capita income of the Indian farm households, as well as the backwardness of the rural financial institutions during the study period, such an outcome is not unexpected. What stands out in the case of India compared with Japan and Taiwan is the lack of net current official transfers as a significant source of surplus outflow from agriculture. This was due to the extremely low tax burden on the agricultural incomes relative to the incomes generated in other sectors of the economy, and particularly the decline in the tax burden in the latter half of the 1960s, when agricultural incomes in real terms were rising due to the terms of trade effect. In this respect, the experience of India is similar to that of China in the post-1979 reform era and to that of Iran.

The Case of Iran

An important feature of the Iranian economy, which distinguishes it from the other countries in our sample, was the substantial inflow of external finance in the form of government oil-export revenues. This provided ample resources which supplemented the investible funds mobilized from the domestic non-oil economy. The availability of external resources also facilitated a process of rapid structural change, and particularly fast income growth in the non-agricultural sector, which contributed to the formation of the popular belief that there was a resource squeeze from the agricultural sector. However, as shown in Chapter 7, there seems to have been a considerable net resource flow into Iranian agriculture which particularly intensified during the 1970s' oil boom. In fact, all the financial mechanisms of surplus transfer, both visible and invisible, recorded a net inflow into the farm sector.[3]

An important source of surplus inflow into Iranian agriculture during the study period on the invisible side was the terms of trade effect. There was a sustained terms of trade improvement in favour of agriculture over the study period which particularly accelerated during the oil boom years of the 1970s. As noted in Chapter 7, the income gains through the terms of trade effect by the end of the study period were more than the real income gains arising from the normal growth of value added in the sector, measured at prices prevailing in the early 1960s. Even measured at base point prices prevailing at the world level, the agricultural terms of trade in Iran seemed to indicate an income gain to the agricultural sector for most

[3] Private bank capital flow, during the 1968–72 period, which showed a relatively small net positive outflow was of course an exception. The negative sign for 'Other' capital in Table 11.1 refers to the period under study excepting the 1968–72 period. For more details, see Ch. 7.

of the study period. This formed one mechanism for transferring part of productivity gains in other sectors as well as the increasing oil export revenues to the agricultural sector.

The income gains through the terms of trade effect were supplemented by other visible inflows on the financial side of the resource flow accounts. An important source of net resource inflow into the farm sector was through the net factor income flow, which mainly consisted of wage-labour income of members of the farm household working in the non-agricultural sector. The rapid rate of growth of the non-agricultural sector meant that this source of income was large enough to overshadow factor income outflows such as interest on loans and rent payments to absentee landlords. With the implementation of the land-reform programme, the latter had lost its significance as a source of surplus outflow from agriculture during the study period.

As with the other countries discussed above, on the capital account the net official flows showed an inflow into Iranian agriculture, which largely reflected the importance of government investment in the sector. This was a particularly large net inflow compared with other sources of resource transfer in the 1970s, as the government, benefiting from large oil price increases, undertook extensive investment projects in the agricultural sector. In addition, the government provided ample subsidized credit through the banking system to the farm sector. In the absence of a matching outflow of finance in the form of private financial investment by the farm households, this meant that private capital flows also registered a net financial inflow into the agricultural sector. This aspect of intersectoral resource flows in Iran is in sharp contrast to agricultural surplus countries (Japan and post-colonial Taiwan), where private capital outflows in the form of financial investment of the farm households formed an important mechanism of surplus outflow from agriculture. Another important feature of resource flows in Iran, which again contrasts sharply with these latter two countries, was the virtual absence of any government taxation of agriculture. As noted above, in the case of countries such as Japan and Taiwan, in the early stages of their development, when rural financial markets were still undeveloped, government taxation formed an important source of surplus outflow from agriculture.

The growing net financial burden of Iranian agriculture during the study period was, of course, connected to the developments in the economy on the real side. Substantial government subsidies on new inputs and credit, as well as the government's direct investment in the agricultural sector, implied a sizeable and fast-growing inflow of new intermediate and capital goods inputs into the sector. The use of new inputs such as chemical fertilizers and pesticides, as well as agricultural machinery and implements, increased at a phenomenal rate, particularly accelerating during the 1970s. The rapid growth of the gross inflow of new inputs into the agricultural sector in the case of Iran was not unlike the experience of Japan and Taiwan, where agriculture made a positive net finance contribution to the other sectors of the economy. What was distinct in the case of Iran, however, was the lack of commensurate response of agricultural value added to the inflow of new producer goods resulting from production and allocative inefficiencies within the agricultural sector. This was an important factor in turning agriculture into a net financial burden on the rest of the economy, as under these circumstances the marketed surplus grew at a slower rate than the purchases of the agricultural sector from the other sectors. This effect was, of course, exacerbated on the consumption side as, due to the sizeable income gains in agriculture through the terms of trade effect and factor income flows, the rate of growth of purchases of consumer goods by the farm households surpassed those of output and marketed surplus of the sector.

11.3 Agricultural Surplus and Industrial Accumulation

The variety of experiences in terms of the direction, relative magnitude, and the patterns and mechanisms of intersectoral resource flows in the five country case studies reveals interesting results regarding different aspects of the role of agriculture in the industrialization process. In Chapter 10 we examined some of the implications of these empirical findings for the general theoretical debate in the literature on the role of agriculture in economic development. An important part of this debate has been on the financing role of agriculture in the process of early industrial accumulation. Our empirical results suggest that, at least in relation to the surplus labour economies studied here, there does not seem to be any significant relation between agricultural surplus and the overall savings and investment rates in the economy. In fact, in China, India, and Iran, where agriculture imposed a large and growing financial burden on the rest of the economy, higher rates of national savings and investment were achieved in a shorter time than in countries such as Japan and Taiwan, where agriculture provided a positive net finance contribution. Furthermore, in the latter two countries, even during the periods when the net finance contribution of agriculture was growing rapidly, the significance of agricultural surplus in the overall national savings declined over time. On the basis of the experience of the countries under study here, therefore, it appears that the role of agriculture in financing industrial accumulation has been historically less significant than suggested in the development literature. In fact, in China, India, and Iran, the non-agricultural sector not only managed to finance high rates of industrial investment, but also generated sufficient savings to finance a growing net financial burden by the agricultural sector.

The empirical evidence discussed in Chapter 10 also indicates that there seems to be no direct relation between net agricultural surplus flow and the growth of agricultural output. The growth of agricultural output in each country was certainly predicated upon a *gross* inflow of new producer and consumer goods from outside the sector on an increasing scale. Whether the final outcome turned out to be a positive or a negative *net* flow of resources into the sector depended primarily on the agricultural output response to the use of the new inputs. In Taiwan, and in

Japan prior to the 1920s, with the relatively high levels of production efficiency and fast rates of productivity growth in the agricultural sector there was a net positive resource outflow from agriculture, despite the rapid growth of agricultural purchases from outside the sector. On the other hand, in China, India, and Iran, agriculture became a growing financial drain on the rest of the economy as a result of the relatively low levels of production efficiency and stagnant agricultural productivity.

There is no doubt that other factors such as the existence of appropriate financial mechanisms of resource flow should be in place for the productivity growth in agriculture to be translated into a net finance contribution by the agricultural sector. A notable example of the failure of the financial mechanisms of resource flow was in the post-1979 reform period in China, where agriculture continued to be a net financial burden on the rest of the economy on an increasing scale despite rapid rates of productivity growth in the sector. In the long run, however, productivity growth seems to be a *sine qua non* for preventing agriculture from turning into a growing financial drain on the rest of the economy. The experience of China during the Maoist period is a forceful reminder of this point. Despite the existence of the institutions of central planning, which were ideal mechanisms for the mobilization and transfer of resources out of agriculture during that period, low productivity growth within the sector entailed a growing financial burden by agriculture on the rest of the economy. On the one hand, low productivity of resource use within the sector meant that the government had to divert a massive amount of new producer goods at highly subsidized prices to agriculture in an attempt to maintain the growth of agricultural output. On the other hand, with the growing population pressure on land and declining productivity of labour in the agricultural sector, the other sectors had increasingly to subsidize the consumption of agricultural households. As a consequence, the institutions of central planning in the agricultural sector, which are widely believed to have been constructed as a means of mobilizing agricultural surplus for industrial accumulation, had instead to be used to secure a net transfer of resources into the agricultural sector.

The experience of Japan provides a further example of the crucial link between the growth of agricultural productivity and net agricultural surplus flow. With the slow-down in productivity growth in Japanese agriculture during the 1920s and 1930s, agricultural surplus began a declining trend until by the end of the study period the balance of net intersectoral resource flows even registered a negative outflow from agriculture. The financial channels of surplus flow, which during the fast productivity growth period prior to the 1920s contributed to the outflow of relatively substantial surplus from agriculture, during the slow productivity phase turned into mechanisms of income transfer to agriculture.

The central role of productive efficiency and productivity growth in agriculture for the process of intersectoral resource flows has already been highlighted above in Chapter 10 and in various country case study chapters and we need not labour this point here any further. The productivity of resource use within different sectors also provides the key to understanding the apparently paradoxical result that in countries such as China, India, and Iran, where strong import-substituting industrialization policies have been followed, the finance contribution of agriculture to other sectors has been negative, in contrast to Taiwan, where agriculture has provided a sizeable net finance contribution to other sectors. This is contrary to the popular belief that import-substitution industrialization is tantamount to a resource squeeze from agriculture. The theoretical underpinning of such a belief, discussed in Chapter 1, is provided by the full-employment, general-equilibrium models of neoclassical theory – in which context the above empirical finding is indeed paradoxical. Within that framework, sectoral outputs become substitutes, as different sectors with given technologies of production compete for fully employed resources, and policies introduced to accelerate the rate of investment in industry would necessarily be at the expense of agricultural investment. As Kaldor (1975, p. 203) has observed: 'this [general equilibrium] approach ignores the essential complementarity between different factors of production (such as capital and labour) and different types of activities (such as that between primary, secondary and tertiary sectors of the economy) which is far more important for the understanding of the laws of change and development of the economy than the substitution aspect'.

As noted in Chapter 1, post-war development economics, by assuming increasing returns in industry and surplus agricultural labour, reintroduced the complementary aspect of sectoral activities at the centre of the development process. Investment in industry makes fuller utilization of surplus labour and thus increases the efficiency of resource use in the economy as a whole. At the same time such investment increases the supply of new land-augmenting producer goods to the agricultural sector – essential for the growth of agricultural output.[4] Agricultural growth would in turn enhance the possibilities for further profitable investment in industry as well as improving industrial productivity by expanding the market for industrial products. As shown in Chapter 9, the massive inflow of producer goods into Chinese agriculture during the 1960s and 1970s was made possible only by a prior investment in industrial activities supplying the agricultural producer goods. This was true also for the other countries in our sample, though the degree to which the agricultural producer goods were supplied by domestic industry varied amongst them depending on the openness of their economies and the orientation of their industrial investment. In none of the countries under study were the resources squeezed out of agriculture because of industrial policies and priorities, and in cases where such an interpretation could possibly be supported (e.g. the early Meiji Japan and Taiwan) this phenomenon was associated with agricultural growth and dynamism rather than stagnation.

[4] The agricultural producer goods need not necessarily be produced by the domestic industry. The surplus labour could be employed in the export oriented industries, providing the foreign exchange for the purchase of agricultural producer goods in the international market.

By focusing on the dynamic and complementary aspects of sectoral interactions in the development process, post-war development economics highlighted the significance of sectoral analysis in the study of economic development. However, as in the rest of mainstream economics, little attention was paid by the development economists to factors which governed the production efficiency within the sectors and activities. This was unfortunate because in later years the problems of poor agricultural performance in countries which followed the industrialization policies advocated by post-war development economics came to be automatically associated with the squeeze of resources out of the agricultural sector without questioning the productivity of resources deployed in the sector. This has sometimes even encouraged the prescription of seemingly easy and even simplistic, if not erroneous, policies to deal with the apparent developmental problems of Third World agriculture. If the root of the problem is seen to be the industrialization policies followed by the governments, then the abandonment of such policies and the reliance on the free play of market forces is said to restore agricultural growth. The mechanism through which this process is supposed to take place is the reallocation of resources away from industry and towards resource-hungry agriculture, which allegedly has been hitherto squeezed due to the industrialization bias of government policy.[5] The result of this simplistic approach has been the neglect of some of the most fundamental forces in the development process, namely, production efficiency and the level and rate of change in productivity of resource use in different sectors.

As noted above, the problems of agricultural development in China, India, and Iran did not arise because of lack of investment and the diversion of resources away from agriculture; on the contrary, agriculture in these countries benefited from a relatively much higher rate of inflow of resources, both on a gross and net basis, than Japan and Taiwan. What distinguished the latter two countries was rather the relatively much higher production efficiency and faster rates of growth in agricultural production. With fast rates of productivity growth, relatively sizeable net resource outflows from agriculture could take place in these two countries through government taxation and adverse terms of trade movement, without apparently adversely affecting the relative rate of return on investment and production incentives in their agricultural sectors. On the other hand, in countries with slow agricultural productivity growth, despite generous tax exemptions and large income transfers into agriculture through favourable terms of trade movements and other mechanisms, agriculture continued to be a major constraint on economic growth.

It would of course be a mistake to make too many generalizations on the basis of

[5] This line of argument is particularly favoured by the World Bank, as evidenced by the title of a recent World Bank publication, *The Plundering of Agriculture in Developing Countries* (Schiff and Valdes, 1992*a*). See also Schiff and Valdes (1992*b*), where this argument is summed up in the preface by Theodore Schultz in the following words: 'the modernization of agriculture was being sacrificed at the altar of industrialization'.

the experience of the five countries studied here. There have been, no doubt, countries where agricultural output has suffered due to lack of investment and a general neglect by the government which may have even led to a resource squeeze out of the sector. The case studies in this book, however, establish that such a resource squeeze need not be a necessary by-product of development strategies favouring industrial growth. If anything, industrialization seems to be a necessary condition for the rapid growth of inflow of new producer goods into agriculture, at least for the surplus labour economies of the type studied here. An even more important point highlighted by the present case studies is that, from a long-term perspective, the efficiency of resource use within the sectors, as determined by organizational and technological capabilities of the different economies, seems to play a more crucial role than sectoral allocation of resources in explaining the different growth performances in the countries studied. Too much emphasis on the question of allocation of resources between sectors, with a focus on the substitution aspects of sectoral interactions, has led to a neglect of the more dynamic considerations related to the efficiency of resource use, notably connected to the process of technological progress. It turns out that the critical question in the study of the long-term comparative performance of agricultural economies in Japan and Taiwan, on the one hand, and in India, China, and Iran, on the other, is related more to the organizational and institutional relations which determined the efficiency of resource use within these respective economies than to the quantity of resources invested in them.[6]

[6] It is interesting that, while in other fields in economics a rich body of literature has developed in the past two decades on the functioning of economic institutions and their implications for production efficiency, the domination of the issues related to resource allocation between activities in the policy debates in development economics has kept it almost insulated from these recent developments. See Nabli and Nugent (1989).

Bibliography

Addison, T., and Demery, L. (1987), 'Stabilization policy and income distribution in developing countries', *World Development*, 15/12 (Dec.), 1483–98.

Agarwala, R. (1983), 'Price distortions and growth in developing countries', World Bank Staff Working Paper 575 (Washington, DC).

Aghazadeh, E., and Evans, D. (1985), *Price Distortions, Efficiency and Growth* (University of Sussex: Institute of Development Studies).

Agrostat (1991), *Agricultural Statistics Package* (Food and Agriculture Organization, Rome).

Ahluwalia, M. S. (1985), 'Rural poverty, agricultural production and prices: A re-examination', in Mellor and Desai (1985), 59–76.

Ahmed, Iftikhar, and Ruttan, Vernon W. (1988), *Generation and Diffusion of Agricultural Factors: The Role of Institutional Factors* (Aldershot).

Anderson, K. (1983), 'Growth of agricultural protection in East Asia', *Food Policy*, 8: 327–36.

Arrow, K. J. (1962), 'The economic implications of learning by doing', *Review of Economic Studies*, 29: 155–73.

Ash, R. F. (1991), 'The peasant and the state', *The China Quarterly*, 127 (Sept.), 493–526.

Ashraf, A. (1982), 'Dehghanan, Zamin va Enghelab', in *Ketab-e Agah*, (Tehran).

——and Banuazizi, A. (1980), 'Policies and strategies of land reform in Iran', in Inayatollah, B. (ed.), *Land Reform: Some Asian Experiments* (Asian and Pacific Administration Center, Kuala Lumpur, Malaysia), 15–60.

Aubert, C. (1988), 'China's food take-off?', in Feuchtwang *et al.* (1988), 101–29.

Bagchi, A. K. (1975), 'Some characteristics of industrial growth in India', *Economic and Political Weekly*, Feb. Annual No.

Balassa, B. (1988), *Quantitative Appraisal of Adjustment Lending*, World Bank Research Working Paper (Washington).

Bank Markazi, Iran (1962–77), *Annual Report and Balance Sheet* (Tehran).

Bardhan, P. K. (1974), 'The pattern of income distribution in India: A review', in T. N. Srinivasan and P. K. Bardhan (eds.), *Poverty and Income Distribution in India* (Statistical Publishing Society, Calcutta), 103–37.

——(1979), 'Wages and unemployment in a poor agrarian economy: A theoretical and empirical analysis', *Journal of Political Economy*, 87: 479–500.

——(1985), 'Poverty and trickle-down in rural India: A quantitative analysis', in Mellor and Desai (1985), 76–95.

Barker, R., and Winkelmann, D. (1974), 'Cereal grains: Future directions for technological change', in Islam (1974).

——Sinha, R., and Rose, B. (eds.) (1982), *The Chinese Agricultural Economy* (Boulder, Col., and London).

Bell, Clive (1979), 'The behaviour of a dual economy under different "closing rules"', *Journal of Development Economics*, 6: 47–72.

Bell, Clive and Srinivasan, T. N. (1984), 'On the uses and abuses of economy-wide models in development policy analysis', *Economic Structure and Performance: Essays in Honour of Hollis B. Chenery* (London), 451–76.

Bharadwaj, Krishna (1979), 'Towards a macroeconomic framework for a developing economy: The Indian case', *The Manchester School* (Sept.), 270–302.

Boyce, James (1987), *Agrarian Impasse in Bengal: Institutional Constraints to Technological Change* (New York).

Byres, T. (1972), 'The dialectic of India's green revolution', *South Asian Review* (Jan).

——(1979), 'Of neo-populist pipe-dreams: Daedalus in the Third World and the myth of urban bias', *Journal of Peasant Studies*, (Jan.), 210–44.

Central Statistical Office (CSO) (1963–77), *Household Budget Survey of Rural Areas* (Tehran).

——(1966–80), *Statistical Yearbook of Iran* (Tehran).

——(1968–77), *Household Budget Survey of Urban Areas* (Tehran).

Chakravarty, S. (1987), *Development Planning: The Indian Experience* (Oxford).

Chaudhary, K. M., and Maharaja, M. (1966), 'Acceptance of improved practices and their diffusion among wheat-growers in the Pali district of Rajasthan', *Indian Journal of Agricultural Economics* (Jan.–Mar.): 161–5.

Chaudhuri, P. (ed.) (1978), *The Indian Economy: Poverty and Development* (London).

——(1971), *Aspects of Indian Economic Development* (London).

Chenery, H. B. (1960), 'Patterns of industrial growth', *American Economic Review*, 50: 624–54.

——(1986), 'Growth and transformation', in H. B. Chenery, S. Robinson, and M. Syrquin (eds.), *Industrialization and Growth: A Comparative Study* (Washington, DC).

——and Srinivasan, T. (eds.) (1988), *Handbook of Development Economics*, i (Amsterdam).

———and Srinivasan, T. (eds.) (1989), *Handbook of Development Economics*, ii, (Amsterdam).

——and Syrquin, M. (1975), *Patterns of Development, 1950–70* (London).

Chichilnisky, G., and Taylor, L. (1980), 'Agriculture and the rest of the economy: macro-connections and policy constraints', *American Journal of Agricultural Economics* (May), 303–9.

Cripps, F. *et al.* (1973), *Growth in Advanced Capitalist Countries* (Cambridge).

Croll, E. J. (1988), 'The new peasant economy in China', in Feuchtwang *et al.* (1988), 77–99.

Dantwala, M. L. (1985), 'Technology, growth, and equity in agriculture', in Mellor and Desai (1985), 110–24.

Denison, E. F. (1962), *The Sources of Economic Growth in the United States and the Alternatives before Us* (Committee for Economic Development, New York).

——(1967), *Why Growth Rates Differ: Post-war Experience in Nine Western Countries* (Washington, DC).

Denman, D. R. (1973), *The King's Vista: A Land Reform which Has Changed the Face of Persia* (Berkhamsted).

Desai, D. K., and Sharma, B. M. (1966), 'Technological change and the rate of diffusion', *Indian Journal of Agricultural Economics* (Jan.–Mar.).

Diamond, M. (1978), 'Towards a change in the economic paradigm through the experience of developing countries', *Journal of Development Economics*, 5: 19–53.

Dixit, A. (1969), 'Theories of the dual economy: A survey', Technical Report 30 (Economic Growth Project, University of California, Berkeley).
——(1973), 'Models of dual economies', in J. A. Mirrlees and N. H. Stern (eds.), *Essays in the Theory of Economic Growth* (New York).
Dobb, M. (1964), 'Some reflections on the theory of investment planning and economic growth', in Polish Scientific Publishers, *Problems of Economic Dynamics and Planning: Essays in Honour of Michal Kalecki* (Warsaw).
Dutt, A. (1984), 'Stagnation, income distribution and monopoly power', *Cambridge Journal of Economics*, 8: 25–40.
——(1989), *Sectoral Balance: A Survey* (Helsinki, World Institute for Development Economics Research).
Ellman, M. (1975), 'Did the agricultural surplus provide the resources for the increase in investment in USSR during the first five-year plan', *Economic Journal*, 85: 844–64.
——(1992), *Collectivisation and Soviet Investment in 1928–32 Revisited* (University of Amsterdam, Faculty of Economics and Econometrics, mimeo).
Fan, S. (1991), 'Effects of technological change and institutional reform on production growth in Chinese agriculture', *American Journal of Agricultural Economics* (May), 266–75
Fei, J. C. H. and Ranis, G. (1964), *Development of the Labor Surplus Economy* (Homewood, Ill.).
———(1966), 'Agrarianism, dualism and economic development', in I. Adelman and E. Thorbecke (eds.), *Theory and Design of Economic Development* (Baltimore).
———and Kuo, S. W. Y. (1978), 'Growth and family distribution of income by factor components', *Quarterly Journal of Economics* (Feb.), 17–53.
————(1979), *Growth with Equity: The Taiwan Case* (Oxford).
Feuchtwang, S., Hussain, A., and Pairault, T. (1988), *Transforming China's Economy in the Eighties*, i. *The Rural Sector, Welfare and Employment* (London).
Fields, G. (1980), *Poverty, Inequality and Development* (New York).
Galenson, W. (1979), *The Economic Development of Taiwan, 1945–75: The Experience of the Republic of China* (Ithaca, NY).
Gandhi, V. P. (1966), *Tax Burden on Indian Agriculture* (Harvard University, Cambridge, Mass.).
Ghosh, J. (1988), 'Intersectoral terms of trade, agricultural growth, and the pattern of demand', *Social Scientist*, 16/4 (Apr./May): 9–27
Griffin, K. (1974), *The Political Economy of Agrarian Change: An Essay on the Green Revolution* (Harvard, Cambridge, Mass.).
——and Ghose, A. K. (1979), 'Growth and impoverishment in the rural areas of Asia', *World Development*, (Apr./May), 361–83.
Grilli, E. R., and Yang, M. C. (1988), 'Primary commodity prices, manufactured goods prices, and the terms of trade of developing countries: What the long run shows', *The World Bank Economic Review*, 2/1: 1–47.
Gutmann, P. (1981), 'The measurement of terms of trade effects', *Review of Income and Wealth*, 27: 433–53.
Hayami, Y., and Ruttan, V. W. (1971), *Agricultural Development: An International Perspective* (Baltimore and London).

Hayami, Y. *et al.* (1975), *A Century of Agricultural Growth in Japan: Its Relevance to Asian Development* (Tokyo).

Helleiner, G. K. (ed.) (1992), *Trade Policy, Industrialization, and Development: New Perspectives* (Oxford).

Hirashima, M. (1985), 'Poverty as a generation's problem: A note on the Japanese experience', in Mellor and Desai (1985).

Hirschman, A. O. (1958), *The Strategy of Economic Development* (New Haven, Conn.).

Ho, S. P. S. (1978), *Economic Development of Taiwan, 1860–1970* (New Haven, Conn.).

Hornby, J. M. (1968), 'Investment and trade policy in the dual economy', *Economic Journal* 78/1: 96–107.

Hwa, E. C. (1989), 'The contribution of agriculture to economic growth: some empirical evidence', in Williamson and Panchamukhi (1989).

International Labour Organization (ILO) (1972), *Employment and Income Policies for Iran, 1972* (Geneva).

Ishikawa, S. (1967a), *Economic Development in Asian Perspective* (Tokyo).

——(1967*b*), 'Resource flow between agriculture and industry: The Chinese experience', *The Developing Economies*, 5/1: 3–49.

——(1988), 'Patterns and processes of intersectoral resource flows: Comparison of cases in Asia', in G. Ranis and T. P. Schultz (eds.), *The State of Development Economics* (Oxford).

Islam, N. (ed.) (1974), *Agricultural Policy in Developing Countries* (New York).

Johnston, B. F. (1966), 'Agriculture and economic dvelopment: The relevance of Japanese experience', *Food Research Institute Studies*, 6: 251–312.

——and Mellor, J. (1961), 'The role of agriculture in economic development', *American Economic Review*, 51/4 (Sept.), 566–93.

Jorgenson, D. W. (1961), 'The development of a dual economy', *Economic Journal*, 71: 309–34.

Kahlon, A. S., and D. S. Tyagi (1980), 'Intersectoral terms of trade', *Economic and Political Weekly*, (Dec.), A173–84.

Kakwani, N. (1980), *Income Inequality and Poverty: Methods of Estimation and Policy Applications* (Oxford).

Kaldor, N. (1962), 'The role of taxation in economic development', in N. Kaldor, *Essays in Economic Policy*, i (London).

——(1964), 'Dual exchange rates and economic development', in N. Kaldor, *Essays in Economic Policy*, ii (London), 178–203.

——(1967), *Strategic Factors in Economic Development* (New York).

——(1972*a*), 'Advanced technology in a strategy of development', in N. Kaldor, *Further Essays on Applied Economics* (London).

——(1972*b*), 'Capitalism and industrial development: Some lessons from Britain's experience', in N. Kaldor, *Further Essays on Applied Economics* (London).

——(1975), 'What is wrong with economic theory', *Quarterly Journal of Economics*, 89/3 (Aug.): 347–57.

——(1976), 'Inflation and recession in the world economy', *Economic Journal*, 86: 703–14.

——(1978), 'The irrelevance of equilibrium economics', in N. Kaldor, *Further Essays on Economic Theory* (London).

Kalecki, M. (1970), 'Problems of financing development in a mixed economy', in W. Eltis *et al.* (eds.) *Induction, Growth and Trade* (Oxford).
——(1976), *Essays in Developing Economies* (London).
Kanbur, S. M. R. (1987), 'Structural adjustment, macroeconomic adjustment and poverty: a methodology for analysis', *World Development* 15/12 (Dec.), 1515–26.
——and McIntosh, J. (1988), 'Dual economy models: retrospect and prospect', *Bulletin of Economic Research*, 40/2: 83–113.
Karshenas, M. (1989), *Intersectoral Resource Flows and Development: Lessons of Past Experience*, World Employment Programme Working Paper 99 (Geneva, ILO).
——(1990*a*), 'Oil income, industrialization bias, and the agricultural squeeze hypothesis: new evidence on the experience of Iran', *Journal of Peasant Studies*, 17/2 (Jan.), 245–72.
——(1990*b*), *Oil, State and Industrialization in Iran* (Cambridge).
——(1993), 'Intersectoral resource flows and economic development: Lessons from past experience', in Ajit Singh and Hamid Tabatabai (eds.)(1993), 179–228.
Katouzian, H. (1978), 'Oil versus agriculture: A case study of dual resource depletion in Iran', *Journal of Peasant Studies*, vol 5, No. 3, 347–69.
——(1981), *The Political Economy of Modern Iran: Despotism and Pseudo-Modernism, 1926–1979* (London).
Kendrick, J. (1973), *Post-war Productivity Trends in the United States, 1948–1969* (New York).
Kikuchi, M., and Hayami, Y. (1985), 'Agricultural growth against a land-resource constraint: Japan, Taiwan, Korea and the Philippines', in K. Ohkawa and G. Ranis (eds.), *Japan and the Developing Countries: A Comparative Analysis* (Oxford).
King, B. (1981), *What is a SAM? A Layman's Guide to Social Accounting Matrices* (Washington, DC).
Kojima, R. (1988), 'Agricultural organization: New forms, new contradictions', *The China Quarterly*, 116 (Dec.), 706–35.
Kravis, I. B., *et al.* (1982), *World Product and Income: International Comparisons of Real Gross Product*, (Baltimore).
Krishna, R., and Raychaudhuri, G. S. (1980), *Trends in Rural Savings and Private Capital Formation in India*, World Bank Staff Working Paper 382 (Washington, DC).
Krishnamurthy, J. (1974), 'Structure of the working force of the Indian Union', (Fourth European Conference on South Asian Studies, University of Sussex).
Krugman, Paul (1990), 'Endogenous innovation, international trade, and growth', in Paul Krugman (ed.) *Rethinking International Trade* (Cambridge, Mass.).
Kung, J. K. (1992), 'Food and agriculture in post-reform China: The marketed surplus problem revisited', *Modern China*, 18/2 (Apr.), 138–70.
Kuo, S. W. Y. (1975), 'Income distribution by size in Taiwan area: Changes and causes', in *Income Distribution, Employment and Economic Development in South East and East Asia*, Papers and Proceedings of Joint Seminar by the Japan Economic Research Center and the Council for Asia Manpower Studies, 80–153.
Kurabayashi, Y. (1971), 'The impact of change in terms of trade on a system of national accounts', *The Review of Income and Wealth*, 17/3: 285–99.
Kuznets, S. (1964), 'Economic growth and the contribution of agriculture: Notes on measurement', in C. Eicher and L. Witt (eds.), *Agriculture in Economic Development* (New York).

Kuznets, S. (1966), *Modern Economic Growth* (New Haven, Conn.).

——(1971), *Economic Growth of Nations: Total Output and Production Structure* (Cambridge, Mass.).

——(1979), 'Growth and structural shifts', in W. Galenson (ed.), *Economic Growth and Structural Change in Taiwan: The Post-war Experience of the Republic of China* (Ithaca, NY).

Lardy, N. R. (1983), *Agriculture in China's Modern Economic Development* (Cambridge).

Lee, T. H. (1971), *Intersectoral Capital Flows in the Economic Development of Taiwan 1895–1960* (Ithaca, NY).

——(1974), 'Food supply and population growth in developing countries: A case study of Taiwan', in Islam (1974).

—(1983), *Agriculture and Economic Development in Taiwan*, i, ii, (Taichung).

Leibenstein, H. (1966), 'Allocative efficiency vs. X-efficiency', *American Economic Review*, 56 (June), 392–415.

——(1978), 'On the basic propositions of X-efficiency theory', *American Economic Review*, 68 (May), 328–32.

Lewis, W. A. (1954), 'Economic development with unlimited supplies of labour', *The Manchester School* (May).

Lin, C. Y. (1973), *Industrialization in Taiwan, 1946–72: Trade and Import-Substitution Policies for Developing Countries* (New York).

Lipton, M. (1977), *Why Poor People Stay Poor: Urban Bias in World Development*, (Cambridge, Mass.).

Lucas, Robert E., jun., (1988), 'On the mechanics of economic development', *Journal of Monetary Economics*, 22: 3–22.

Lysy, Frank J. (1980), 'Investment and employment with unlimited labour: The role of aggregate demand', *Journal of Development Economics*, 7: 541–66.

McLachlan, K. S. (1988), *The Neglected Garden: The Politics and Ecology of Agriculture in Iran* (London).

Majd, M. H. (1983), 'Land reform and agricultural policy in Iran, 1962–78', *International Agricultural Economics Study* (New York, Dept. of Agricultural Economics, Cornell University).

Marx, Karl (1976), *Capital*, i (London).

——(1977), *Capital*, iii (London).

Mehran, F. (1975), *Income Distribution in Iran: The Statistics of Inequality*, Working Paper 30 (Geneva, ILO).

Mellor, J. W. (1973), 'Accelerated growth in agricultural production and the intersectoral transfer of resources', *Economic Development and Cultural Change*, 22/1 (Oct.), 1–16.

——(1989), 'Rural employment linkages through agricultural growth: Concepts, issues, and questions', in Williamson and Panchamukhi (1989).

——and Desai, G. M. (eds.) (1985), *Agricultural Change and Rural Poverty: Variations on a Theme by Dharm Narain* (Baltimore).

——*et al.* (1968), *Developing Rural India* (Ithaca, NY).

Millar, J. R. (1970), 'Soviet rapid development and the agricultural surplus hypothesis', *Soviet Studies*, 22: 77–91.

Ministry of Interior Affairs (1960), *National Agricultural Census* (Tehran).

Mitra, A. (1977), *Terms of Trade and Class Relations* (London).

Mody, A. (1981), 'Resource flows between agriculture and non-agriculture in India, 1950–1970', *Economic and Political Weekly*, Mar. Annual No.
——Mundle, S., and Raj, K. N. (1985), 'Resources flows from agriculture: Japan and India', in K. Ohkawa and G. Ranis (eds.), *Japan and the Developing Countries: A Comparative Analysis* (Oxford).
Morrisson, C., and Thorbecke, E. (1990), 'The concept of the agricultural surplus', *World Development*, 18/8: 1081–95.
Mundlak, Y., Danin, Y., and Tropp, Z. (1974), 'Agriculture and economic growth: some comments', in Islam (1974).
Mundle, S. (1981), *Surplus Flows and Growth Imbalances* (New Delhi).
——(1985), 'The agrarian barrier to industrial growth', *Journal of Development Studies* (Oct.).
——and Ohkawa, K. (1979), 'Agricultural surplus flow in Japan, 1888–1937', *The Developing Economies*, 17 (Sept.).
Murphy, K. M., Sheifer, A., and Vishny, R. W. (1989), 'Industrialization and the big push', *Journal of Political Economy*, 97/51: 1003–26.
Nabli, M. K., and Nugent, J. B. (1989), 'The new institutional economics and its applicability to development', *World Development*, 17/9: 1333–47.
Narain, D. (1961), *Distribution of the Marketed Surplus in Agricultural Produce by Size-level of Holding in India, 1950–51* (Bombay).
——(1976), *Growth of Productivity in Indian Agriculture*, Cornell Occasional Paper 93 (Ithaca, NY).
Nicholls, W. H. (1961), 'Industrialisation, factor markets and agricultural development', *Journal of Political Economy*, 69 (Aug.).
——(1963), 'An agricultural surplus as a factor in economic development', *Journal of Political Economy*, 71.
Nishimizu, M., and Page, John M., jun. (1989), 'Productivity change and growth in industry and agriculture', in Williamson and Panchamukhi (1989), 390–413.
Nowshirwani, V. F. (1976), *Technology and Employment Programme: Agricultural Mechanisation in Iran*, Working Paper 28 (Geneva, ILO).
Nurkse, R. (1953), *Problems of Capital Formation in Underdeveloped Countries* (Oxford).
Ohkawa, K. (1979), 'Aggregate growth and product allocation', in Ohkawa, Shinohara, and Meissner (1979).
——and Rosovsky, H. (1960), 'The role of agriculture in modern Japanese economic development', *Economic Development and Cultural Change*, 9: 43–67.
——and Shinohara, M. (eds., with L. Meissner) (1979), *Patterns of Japanese Economic Development: A Quantitative Appraisal* (New Haven, Conn.).
——Shimizu, Y., and Takamatsu, N. (1978), 'Agricultural surplus in an overall performance of savings-investment', in *Japan's Historical Experience and the Contemporary Developing Countries: Issues for Comparative Analysis – Papers and Proceedings of the Conference* (Tokyo, International Development Center of Japan).
————(1982), *'Agricultural surplus' in Japan's case: Implications for Various Possible Patterns in the Initial Phase of Development* (Tokyo, International Development Center of Japan).
Oshima, H. (1973), 'Income distribution', Mission Working Paper 11 (Geneva, ILO).

Oshima, H. (1986), 'The transition from an agricultural to an industrial economy in East Asia', *Economic Development and Cultural Change*, 34/4: 783–809.
Owen, F. (1966), 'The double development squeeze on agriculture', *American Economic Review*, 56/1 (Mar.): 43–70.
Pang, C. (1992), *The State and Economic Transformation* (New York).
Passin, H. (1965), *Society and Education in Japan* (New York).
Patnaik, U. (1975), 'Contribution to the output and marketable surplus of agricultural products by cultivating groups in India, 1960–61', *Economic and Political Weekly* (Dec.): A90–110.
——(1988), 'Some aspects of development in the agrarian sector', *Social Scientist*, 16/2 (Feb.): 17–40.
Perkins, D., and Yusuf, S. (1984), *Rural Development in China* (Baltimore).
Pesaran, M. H. (1976), 'Income distribution and its major determinants in Iran', in J. W. Jacqz (ed.), *Iran: Past, Present and Future* (Colombia), 267–86.
——and Ghahvary, F. (1978), 'Growth and income distribution in Iran', in R. Stone and W. Peterson (eds.), *Econometric Contribution to Public Policy* (London).
Preobrazhensky, E. A. (1980), *The Crisis of Soviet Industrialisation: Selected Essays*, D. A. Filtzer (ed.) (London).
Pyatt, G., and Thorbecke, E. (1976), *Planning Techniques for a Better Future* (Geneva, ILO).
——and Roe, A. (1977), *Social Accounting for Development Planning with Special Reference to Sri Lanka* (Cambridge).
Quesnay, Francois (1758), *Tableau èconomique*, Eng. trans. (London).
Rada, E. L., and Lee, T. H. (1963), 'Irrigation investment in Japan', Joint Commission on Rural Reconstruction, Economic Digest Series 14.
Radwan, S. (1975), *Employment Implications of Capital Intensive Industries in Iran* (Geneva, ILO).
Raghavan, S. N. (1984), *Report on Impact of Agricultural Investment* (Rome, 1984).
Raj, K. N. (1976), 'Output and stagnation of Indian industries', *Economic and Political Weekly*, Feb. Annual No.
Rakshit, M. (1982), *The Labour Surplus Economy* (Delhi).
Rangarajan, C. (1982), *Agricultural Growth and Industrial Performance in India*, Research Report 33, International Food Policy Research Institute.
Rao, C. H. H. (1975), *Technological Change and the Distribution of Gains in Indian Agriculture* (Delhi).
Raquibuz Zaman, M., and Bose, S. R. (1974), 'Extension service, education and agricultural development, with special reference to Bangladesh', in Islam (1974).
Ricardo, D. (1951), *On the Principles of Political Economy and Taxation*, i. *The Works and Correspondence of David Ricardo*, ed. Piero Sraffa (Cambridge).
Romer, P. M. (1986), 'Increasing returns and long-run growth', *Journal of Political Economy* 94 (Oct.), 1002–37.
——(1990), 'Endogenous technical change', *Journal of Political Economy*, 98: S71–102.
Rosenberg, N. (1976), *Perspectives on Technology* (Cambridge).
Rosenstein-Rodan, P. N. (1943), 'Problems of industrialisation in Eastern and South-Eastern Europe', *Economic Journal* 53: 202–11.

Sah, R. K., and Stiglitz, J. (1984), 'The economics of price scissors', *American Economic Review*, 74: 125–38.
———(1987), 'Price scissors and the structure of the economy', *Quarterly Journal of Economics*, 102/1 (Feb.), 109–34.
Saith, A. (1981), 'Production, prices and poverty in rural India', *Journal of Development Studies*, 17 (Jan.), 16–23.
Salehi-Isfahani, D. (1989), 'The political economy of credit subsidy in Iran, 1973–1978', *International Journal of Middle Eastern Studies*, 21: 359–79.
Sanghvi, P. (1969), *Surplus Manpower in Agriculture and Economic Development With Special Reference to India* (London).
Schiff, M., and Valdes, A. (1992*a*), *The Plundering of Agriculture in Developing Countries* (Washington).
———(1992*b*), *The Political Economy of Agricultural Pricing Policy: A Synthesis of the Economics in Developing Countries* (Baltimore).
——(1971), 'The aspects of Indian education', in Chaudhuri (1971).
Sen, A. (1981), 'The agricultural constraint to economic growth: The case of India', Ph.D. thesis (Cambridge).
——(1986), 'Shocks and instabilities in an agriculture-constrained economy, India 1964–85', *Social Scientist*, 161 (Oct.): 27–48.
Sen, A. K. (1957), 'Some notes on the choice of capital intensity in development planning', *Quarterly Journal of Economics*, 71 (Nov.): 561–84.
Seton, F. (1957), 'The "Transformation Problem"', *Review of Economic Studies*, 24: 149–60.
Shafa-eddin, S. M. (1980), 'A critique of development policies based on oil revenues in recent years in Iran', Ph.D. thesis (Oxford).
Sheng, Y. (1992), *Intersectoral Resource Flows and China's Economic Development* (London).
Shetty, N. S. (1968), 'Inter-farm rates of technological diffusion in Indian agriculture', *Indian Journal of Agricultural Economics* (Jan.–Mar.).
——(1971), 'An inter-sectoral analysis of tax burden and taxable capacity', *Indian Journal of Agricultural Economics* (July–Sept.).
——(1978), 'Structural retrogression in the Indian economy since the mid-sixties', *Economic and Political Weekly*, Feb. Annual No. 185–244.
Shinohara, M. (1970), *Structural Changes in Japan's Economic Development* (Tokyo).
Singh, A., and Tabatabai, H. (eds.) (1993), *The World Economic Crisis and Third World Agriculture: The Changing Role of Agriculture in Economic Development* (Cambridge).
Solow, R. M. (1957), 'Technical change and the aggregate production function', *Review of Economics and Statistics*, 31: 312–20.
Srinivasan, T. N. (1979), 'Trends in agriculture in India, 1949/50–1977/78', *Economic and Political Weekly*, Aug. Special No. 1283–94.
State Statistical Bureau (1987, 1991), see *Statistical Yearbook of China.*
Statistical Yearbook of China (1987) (Hong Kong).
——(1991) (Hong Kong).
Stavis, B. (1982), 'Rural institutions in China', in Barker *et al.* (1982).
Stiglitz, J. E. (1989), 'Economic organization, information, and development', in Chenery and Srinivasan (1989), 93–160.

Streeten, P. (1959), 'Unbalanced growth', *Oxford Economic Papers*, 11: 167–90.

Stuvel, G. (1956), 'A new approach to the measurement of terms of trade effects', *The Review of Economics and Statistics* (Aug.) 294–307.

Syrquin, M. (1988), 'Patterns of structural change', in Chenery and Srinivasan (1988), 203–73.

Tang, A. M. (1984), *An Analytical and Empirical Investigation of Agriculture in Mainland China, 1952–1980* (Taipei, Chung-Hua Institution for Economic Research).

Taylor, L. (1983), *Structuralist Macroeconomics* (New York).

——(1985), 'Demand composition, income distribution and growth', Dept. of Economics, Massachusetts Institute of Technology, unpublished.

——and Arida, P. (1988), 'Long-run income distribution and growth', in Chenery and Srinivasan (1988).

——and Bacha, Edmar L. (1976), 'The unequalizing spiral: A first growth model for Belindia', *Quarterly Journal of Economics*, 90: 197–218.

Teranishi, J. (1982), *Nippon no Keizai Hatten to Kinyu* (Tokyo).

Thamarajakshi, R. (1969), 'Intersectoral terms of trade and marketed surplus of agricultural produce, 1951–52 to 1965–66', *Economic and Political Weekly* (June).

Timmer, C. Peter (1988), 'The agricultural transformation', in Chenery and Srinivasan (1988), 275–331.

Tyagi, D. S. (1979), 'Farm prices and class bias in India', *Economic and Political Weekly*, Sept. Review of Agriculture, A111–24.

——(1988), 'Intersectoral terms of trade in India: misconceptions and fairy tales', *Economic and Political Weekly* (Apr.) 858–64.

Umemura, M. (1979), 'Population and labour force', in Ohkawa and Shinohara (1979).

United Nations (1965–87), *Yearbook of Industrial Statistics.*

——(1968), 'A system of national accounts', *Studies in Methods*, F/2.3.

US Congress, Joint Economic Committee (1967), *An Economic Profile of Mainland China* (Washington, DC).

Veblen, T. (1932*a*), 'On the nature of capital', in Veblen (1932*b*), 324–86.

——(ed.) (1932*b*), *The Place of Science in Modern Civilization, and Other Essays* (New York).

Viner, J. (1937), *Studies in the Theory of International Trade* (New York).

Vittal, N. (1986), 'Intersectoral terms of trade in India: A study of concept and method', *Economic and Political Weekly,* Dec. Review of Agriculture, A147–66.

——(1988), 'Intersectoral terms of trade in India: Reality and hype', *Economic and Political Weekly*, Sept. Review of Agriculture, A133–40.

Wiemer, C. (1992), 'Price reform and structural change: Distributional impediments to allocative gains', *Modern China*, 18/2 (Apr.), 171–96.

Williamson, J. G., and Panchamukhi, V. R. (1989), *The Balance Between Industry and Agriculture in Economic Development* (Proceedings of the Eighth World Congress of the International Economic Association) (London).

Wong, C. P. W. (1982), 'Fiscal reform and local industrialization: The problematic sequencing of reform in post-Mao China', *Modern China*, 18/2 (Apr.), 197–227.

World Bank (1983*a*), *China: Socialist Economic Development*, i, ii (Washington, DC).

——(1983*b*), *World Development Report, 1983* (New York).

——(1986), *World Development Report, 1986* (New York).

——(1988), *Report on Adjustment Lending* (Washington, DC).

——(1990), *Adjustment Lending Policies for Sustainable Growth*, (Country Economics Department, Washington).

Yamada, S., and Hayami, Y. (1979*a*), 'Agriculture', in Ohkawa and Shinohara (1979), 85–104.

———(1979b), 'Growth rates of Japanese agriculture, 1880–1970', in Y. Hayami, V. W. Ruttan, and H. M. Southworth (eds.), *Agricultural Growth in Japan, Taiwan, Korea and the Philippines* (Honolulu).

Young, A. (1928), 'Increasing return and economic progress', *Economic Journal*, 38 (Dec.): 527–42.

Index